From ZERO to INFINITY

By the author

From Zero to Infinity
Introduction to Higher Mathematics

Constance Reid

From ZERO
to INFINITY

What Makes
Numbers Interesting

Second Revised Edition

THOMAS Y. CROWELL COMPANY, NEW YORK

A portion of this book first appeared in somewhat different form as an article in the *Scientific American* and is used here with the permission of the editors.

CONTENTS

ZERO   1
ONE   15
TWO   28
THREE   41
FOUR   57
FIVE   71
SIX   83
SEVEN   97
EIGHT   110
NINE   122
. . .   137
INDEX   157

From ZERO to INFINITY

ZERO IS THE FIRST OF TEN SYMBOLS—THE digits—with which we are able to represent any of an infinitude of numbers. Zero is also the first of the numbers which we must represent. Yet zero, first of the digits, was the last to be invented; and zero, first of the numbers, was the last to be discovered.

These two events, the invention and the discovery of zero, tardy as they were in the history of number, did not occur at the same time. The invention of zero preceded its discovery by centuries.

At the time of the birth of Christ, the idea of zero as symbol or number had never occurred to anyone. The problem of writing down numbers without using a different symbol for each one had been met in very much the same way by all the great civilizations. The Egyptians had used appropriate pictures; the Greeks, the letters of their alphabet; the Romans, the few simple lines that we see so often on cornerstones; but all had grouped the numbers so that the same symbols could be used over and over. It was possible to write down numbers, but it was not possible to write them down in a way that they could be easily handled in even the simplest processes of arithmetic. Anyone who has tried to multiply Roman numerals will have no difficulty in understanding why the Roman, when he had a problem in arithmetic, turned his back on the V's and X's and C's and M's of the written numbers and obtained his answer with the beads on a counting board. Egyptian and Greek did the same thing. Yet it never occurred to any of them that in these same beads was the essential of the most efficient method of number representation the world was to develop in the next two thousand years.

The counting board, although it took various forms and names in various civilizations, was basically a frame divided into parallel columns. Each column had the value of a power of ten, the number of times that a particular power occurred in a total being represented by markers of some sort, usually beads. All of the beads were identical in appearance and all stood for one unit. The value of the unit, however, varied with the column. A bead in the first column had the value of one (10^0); a bead in the second, of ten (10^1); in the third, of one hundred (10^2); and so on.* Numbers which the Romans represented in writing as CCXXXIV (234) and CDXXIII (423) were simply represented and easily distinguishable on the board.

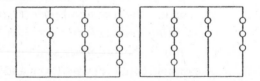

We of today are immediately struck by the resemblance between this ancient method of representing numbers on the counting board and our own method of representing

---

* The uncertain life of a favorite at the court of a tyrant was compared to that of a marker on the counting board "which signifies now much, now little."

them in writing. Instead of nine beads, we use nine different symbols to represent the total of beads in a column and a tenth symbol to indicate when a column is empty. The ordering of these ten symbols, which we call the digits, tells us exactly the same thing the beads do: 234 tell us 2 hundreds, 3 tens, 4 ones, while 423 tells us 4 hundreds, 2 tens, and 3 ones.

In short, modern *positional* notation, where each digit has a varying value depending upon its position in the representation of a number, is simply the notation of the counting board made permanent. All that is needed to transfer a number from the board to paper is ten different symbols; for there can be only one of ten possible totals in a column: one, two, three, four, five, six, seven, eight, or nine beads, or no beads at all. The column can be empty, and the tenth symbol must of necessity be a symbol for such an empty column. Otherwise it would be impossible to distinguish among different numbers from the counting board.

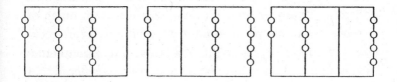

Without such a symbol the above examples would all on paper be the same: 234. With a symbol they are easily distinguishable as 2340, 2034, 2304.

It would seem that the first time anyone wanted to record a number obtained on the counting board, he would automatically have put down a symbol of some sort, a dash, a dot, or a circle, for that empty column—which we today represent by zero. But in thousands of years, nobody did.

3

Not Pythagoras.

Not Euclid.

Not Archimedes.

For the great mystery of zero is that it escaped even the Greeks.†

The respect in which mathematicians hold these ancient "contemporaries" was expressed by England's G. H. Hardy (1877–1947) when he wrote: "Oriental mathematics may be an interesting curiosity, but Greek mathematics is the real thing. As Littlewood said to me once, 'The Greeks are not clever schoolboys or scholarship candidates, but *fellows at another college.*'"

That zero, or nothing, was not recognized as a number by the Greeks is more than curious. They were the first people to be interested in numbers solely because numbers are interesting, and they left to number theory some questions which remain unanswered to this day. They were concerned, however, with learning the secrets of numbers, not with using them; and this may be the reason that the idea of zero as a number escaped them. For although much of number theory has no need for a zero, without a zero reckoning is impossibly hobbled. The great Greek mathematicians, pondering the interesting numbers, considered that reckoning was an occupation for slaves and left it to them.

It was India that gave the world zero, and with it a practical system of arithmetical notation. Sometime in the early centuries of the Christian era, an unknown Hindu who wanted to keep in permanent form the answer on his counting board put down a symbol of his own invention, a

† It is difficult, if not impossible, as the reader of this book will soon discover, to write about numbers without writing about the Greeks.

dot he called *sunya,* to indicate a column in which there were no beads.

Thus, after all the others, came zero, the first of the digits.‡

It must be understood that the dot sunya which the Hindu invented was not the number zero. It was merely a mechanical device to indicate an empty space, and that was what the word itself meant—empty. The Indians still use the same word and symbol for the unknown in an equation—what we usually refer to as *x*—the reason being that until a space is filled with the proper number it is considered to be empty.

With sunya, the symbol zero had been invented; but the number zero was yet to be discovered.

When finally the Indian notation made its way to Europe, it was through the Arabs and as "Arabic" notation. Immensely superior as it was, it was not immediately accepted. Merchants recognized its usefulness while the more conservative class of the universities hung on to the numerals of the Romans and the system of the abacus.§ It was recognized by everyone that the really revolutionary thing about the new notation was the inclusion of the dot —*sifr,* as it was called in Arabic—to represent the empty column. The whole new system came to be identified by the name of this one symbol; and that is how the word *cipher,* in addition to standing for zero, came also to stand

‡ It has been pointed out that the invention of a symbol for nothing, the void, was one for which his philosophy and religion had peculiarly prepared the Hindu.

§ In 1300 the use of the new numerals was forbidden in commercial papers because they could be more easily forged than the Roman numerals. It was not until 1800 that they were completely accepted all over Europe.

for any of the digits and the verb *to cipher*, for *to calculate*.
(*Zero* came later from the Italian.) But sifr, like sunya,
was still a symbol for an empty column, not a number.

Even today, although we use zero constantly, we do not
always think of it as a number. On a typewriter keyboard
or a telephone dial, we list it with the other digits but
place it after nine. Since in value it does not exceed nine,
it is obviously there as a symbol, not as a number.

This should not surprise us, for zero is the one digit which
we do not commonly use as a number. If the reader will
answer the few problems below he will discover for himself that he is much more efficient in handling zero the
symbol than in handling zero the number. The symbol is
the zero he knows; for it is a curious fact that positional
arithmetic, which depends for its existence upon the symbol zero, gets along very well without the number zero.

A TEST

On Zero as a Symbol	*On Zero as a Number*
$1 + 10 =$	$1 + 0 =$
$10 + 1 =$	$0 + 1 =$
$1 - 10 =$	$1 - 0 =$
$10 - 1 =$	$0 - 1 =$
$1 \times 10 =$	$1 \times 0 =$
$10 \times 1 =$	$0 \times 1 =$
$10 \times 10 =$	$0 \times 0 =$
$10 \div 1 =$	$0 \div 1 =$
$1 \div 10 =$	$1 \div 0 =$
$10 \div 10 =$	$0 \div 0 =$

ANSWERS

As a Symbol: 11, 11, −9, 9, 10, 10, 100, 10, 1/10, 1.
As a Number: 1, 1, 1, −1, 0, 0, 0, 0, impossible, indeterminate.

Centuries after sunya had been invented as a symbol for the empty column on the counting board, man was still fumbling toward the mastery of zero as a number which could be added, subtracted, multiplied, and divided like the other numbers. To the scholar of today, poring over ancient mathematical papers, the test of mastery is always the same. Addition, subtraction, even multiplication with zero seem to have caused relatively little trouble. Always it is the handling of division of and by zero that shows us today whether a man really understood the curious new number. The problems that caused the trouble were similar to the last three in our little test (probably the same ones that caused the reader trouble).

$$0 \div 1 =$$

The fractional expression 0/1, which is just another way of expressing the division, is mathematically meaningful. Zero can be divided by any other number; in this it is unique among the numbers. (In number theory, one number is considered to "divide" another only when the answer obtained is a whole number.) No matter what number is multiplied by zero, the answer is always the same—zero. Since $0 \times 1 = 0$, $0 \div 1 = 0$. No matter what number is divided into zero, the answer is always the same—zero.

$$1 \div 0 =$$

The expression 1/0 is not, on the other hand, mathematically meaningful. Zero cannot divide any number except itself, not even as the denominator in a fractional expression. In this, as in the fact that it can be divided by all numbers, it is unique. The reason that 1/0 is a meaningless ex-

7

pression is the same reason that 0/1 is a meaningful one. No matter what number is multiplied by zero, the answer is always zero. A division, however, indicates that some number (the quotient) when multiplied by another (the divisor) will produce the number being divided. If there is an answer to the problem $1 \div 0$, or a value for the expression 1/0, it would have to be such a number that multiplied by zero it would produce one. But we have already stated that *any* number multiplied by zero can produce only zero. It follows, therefore, that we cannot divide one (or any other number) by zero.

$$0 \div 0 =$$

The expression 0/0 is neither mathematically meaningful nor meaningless. It is *indeterminate*. Zero can be divided by itself, but there is no way of determining what the value of the answer is. Since any number multiplied by zero produces zero, zero divided by zero can yield any number. Zero divided by zero can equal zero, since $0 \times 0 = 0$, but it can also equal one, since $0 \times 1 = 0$, and two, since $0 \times 2 = 0$, and so on and on. ||

These three terms we have been using—*meaningful, meaningless,* and *indeterminate*—can perhaps be made clearer by a comparison. An indicated operation of division is said to be mathematically meaningful only if it stands for a specific value that can be obtained by performing the operation. It may be compared to a title used to identify specifically a person not named. The President of the United States, for instance, is such a title. When we use it

|| Zero has always been a favorite in a field of insult which can best be described as "mathematical invective." A recent example appearing in the newspapers is "a lousy nothing divided by nothing." This is mathematically a less definite insult than it was intended to be.

we are referring to a certain person as specifically as if we had named him. In a similar way the expression 0/1 (or $0 \div 1$) refers to a specific value: zero. It cannot stand for another value any more than 10/1 can stand for a value other than ten.

An indicated operation of division, on the other hand, is mathematically meaningless when it cannot possibly have any value. In the same way a title may be meaningless. The King of the United States is such a title. The expression 1/0 (or $1 \div 0$) is meaningless because one cannot be divided by zero; therefore, the expression stands for no value. (No value is not at all the same thing as zero.)

The expression 0/0 (or $0 \div 0$) is meaningless in a quite different sense. It is like the title the United States Senator, which is meaningless for identification unless the context in which it is used tells which of the ninety-six senators is meant by it. The choice with the expression 0/0 is much greater than ninety-six. It can have any numerical value we choose to give it, since any number multiplied by zero produces zero. The expression 0/0 is meaningless only because it can mean anything. Mathematicians say, more technically, that it is *indeterminate,* and it took them centuries to realize that it is. Only then had they finally mastered zero the number.

To understand the special significance of zero among the numbers, we must examine what are known as the integers. When the integers are arranged in order, the positive numbers, which we might say count things present, extend indefinitely to the right; the negative numbers, which count things absent, extend indefinitely to the left. This is an arrangement we are familiar with on the thermometer, the positive numbers being the degrees "above" zero; the negative, those "below."

$$\ldots, -5, -4, -3, -2, -1, 0, +1, +2, +3, +4, +5, \ldots$$

In this arrangement of negative and positive integers, every consecutive pair must be the same distance apart as every other pair. Such regularity of spacing is the essence of the integers: -1 is the same distance from -2 that $+1$ is from $+2$ and also $+2$ from $+3$. But this regularity can be maintained only if zero is included as one of the integers. Without zero the distance between -1 and $+1$ is twice the distance between any other pair. Obviously then -1 and $+1$ are not consecutive: zero is the number between them.

In the Christian accounting of time, unlike on the scale above, zero marks not a number but a point. A problem in degrees of temperature, therefore, yields quite a different answer from a similar problem in years. If the temperature is 5° below zero in the morning and rises 8° during the day it is then 3° above zero. But a child born January 1, 5 B.C., will not be eight years old until A.D. 4. The reason for the difference in the answers to these two apparently identical problems is clear when we place the scale of temperature against the scale of time.

8 DEGREES

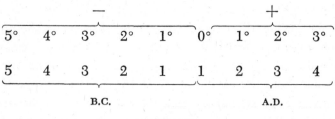

8 YEARS

This difference was the cause of a major howler in the scholarly world in 1930. The celebration of the 2000th anniversary of the birth of the poet Virgil was in full swing when a mathematical killjoy pointed out that there having been no year zero, the poet (born in 70 B.C.) would not be two thousand years old until 1931! The scholars, who should have known better, were performing, on a scale on which zero is not a number, a mathematical process which works as it does because zero is a number.

Among the whole numbers, or integers, zero is unique, being neither negative nor positive. Although we use all of the integers in computation, it is those after zero that we think of as "the numbers." (As late as the twelfth century the Indian mathematician Bhāskara gave $x = 50$ and $x = -5$ as the roots of the equation $x^2 - 45x = 250$, but cautioned, "The second value in this case is not to be taken, for it is inadequate; people do not approve of negative roots.") We even call these numbers that follow zero the *natural* numbers, although it may be argued whether they are in fact more natural than any other numbers. They are the numbers we count with, which seems the natural thing to do with numbers. We do not think of zero as one of them because it does not seem at all natural to most of us to "count" with nothing.#

Yet zero, unlike the negative numbers, is logically at home with the so-called natural numbers, even though it is not positive and they are. For zero answers the same great question that all the counting numbers answer, and it answers it in exactly the same way. The question is simply, *How many?*

An exception may be the proprietor of a shoe store at 0 Newberry Street in Boston.

How many people are there in the room where you are reading this book?

How many elephants are there in the room where you are reading this book?

The answer to the first question is at least one, maybe two or three; but the answer to the second is quite probably zero. The number of elephants in the room is zero. Zero is a number just like one and two and three.

But if zero is a number, the reader may well ask, just what is a number anyway?

Certainly a number is an abstraction, a recognition of the fact that collections may have something in common even though the elements of the collection have in common nothing whatsoever. There is a similarity between two mountains and two birds even though birds and mountains are not similar, and this similarity they share with two of anything. While this may seem obvious to us, it was not to our ancestors. They recognized the difference between one pheasant and two, one day and two, but, as Bertrand Russell has pointed out, "It must have required many ages to discover that a brace of pheasants and a couple of days were both instances of the number two."

What was discovered, mathematically speaking, was that the number two is the common property of all sets containing a pair of anything. It does not matter whether a set contains people, elephants, flies, or mountains, or a miscellaneous collection of the same; it shares with all other sets that contain a pair the number two.

When we say that one, two, and three are numbers, we mean that one is the number of all those sets which

12

contain a single member; two is the number of the sets which contain a pair; and three is the number of the sets which contain three members. Since there is no end to the possibilities of what these sets may contain, we say that they are infinite.

There is a set of zero too, comparable to these others. This is the set which contains no people, no elephants, no flies, no mountains. In other words, the empty set. In the same way that one, two, and three are the numbers of the sets of one, two, and three respectively, zero is the number of the empty set.

There is, however, a difference between the set of zero and the other sets which has nothing to do with the difference in the number of members. While all the other numbers represent an infinite number of sets, zero represents only one, the empty set. Whether it is empty of men, elephants, flies, or mountains does not matter; it is the same set—and there is only one.

It is things like this which make zero a very interesting number among an infinitude of interesting numbers. Each of the natural numbers is, of course, unique: two is not three, and three is not four, and four is not five—or any other number. But the uniqueness of zero is more general than that of the other numbers, more significant for that reason, and therefore more interesting.

Zero is the only number which can be divided by every other number, and the only number which can divide no other number.

Because of these two characteristics, zero is almost invariably a "special case" among the numbers, and we shall find many examples of its "specialness" in the pages to follow. Zero is enough like all the other natural numbers to be one of them, but enough different to be a very interesting number: the last, and the first, of the digits.

A PROBLEM

The digits can be arranged in various ways. In this chapter we have mentioned two. In one arrangement zero follows nine as a symbol; in the other, much less common, zero the number comes before one. But usually, even in fun, zero ends up as the last of the digits. The basis for the arrangement below is one that many mathematicians have great difficulty in perceiving.

8 5 4 9 1 7 6 3 2 0

ANSWER

The digits are arranged in alphabetical order. Secretaries usually outwit mathematicians on this one. An amusing variation is to arrange the digits in a language other than English.

1

We are all familiar with the behavior of the number one in the ordinary processes of arithmetic. It does not surprise us as does the behavior of zero. In fact, it is so simple that we generally dismiss it as trivial. We do not even bother to learn the "ones" in school, so obvious is it to us that any number when multiplied by one yields a product which is itself and when divided, a quotient which is itself. Yet these simple characteristics of the number one have the greatest implications for the study of numbers.

The very first idea of number comes with the recognition that there is a difference between one and more than one. A child grasps this distinction when he is about eighteen months old, and presumably man recognized it proportionately early in his history.*

Either there is one wolf or there are many around the fire, one river or more than one between this camp and the next, one star in the early evening sky or many when the campfire dies down. We begin

* Arnold Gesell writes in *The First Five Years of Life:* "[The 18-month-old infant] has an interest in many and in more. He likes to assemble the many cubes into a pile or to disperse the pile into the many cubes. . . . In comparison, the 1-year-old is single- and serial-minded."

with two number words, but only the number one. Nevertheless it is possible for us to count in a fashion, and quite accurately, with this number. We look up suddenly from the campfire and see "many" wolves—more than one. There are actually two, but we have no word for two, so we say there are many. How many? We try to think of something as "many" as the wolves and come up with another pair with which we are familiar. We announce that there are as many wolves as a bird has wings.

This method of communicating the exact number of wolves, distinguishing among the many meanings of "many," need not stop with a pair. We can find other sets with which we are familiar and against which we can match the number of wolves, one to one. The wings of a bird may be followed by the leaves of a clover, the legs of an animal, the fingers of a hand. We are then able to "count" any number of wolves from one to five although we still have no number other than the number one.

We look around for what is logically the next set, a set which contains one more member than our hand has fingers. It is not so easy to find a set of six in nature. So instead of using another completely new set for counting one more wolf, we add to the set of fingers on one hand a finger from the other. This is a good practical idea because now, for our next set, when the number of wolves increases by still one more, we can add another finger and we can continue in this manner until we have used all the fingers of both hands to count ten wolves.

But the wolves keep coming. What can we do with a many which is more than the fingers on both our hands? We could of course start upon our toes, and some people did; but we decide to reuse our fingers. For one more wolf, we put up both our hands and then a finger by itself. We have now started, inexorably, upon our way to infinity. We will never get there, but we will never have to stop along the way and say that we can't go any farther. For no matter how many wolves we have "counted" and how many fingers we have used in the process, we can always lift one more and count one more wolf.

What then has been our achievement?

Simply that we have constructed with no other number concept except that of the number one, the infinite set of natural numbers.

$$1$$
$$1 + 1$$
$$1 + 1 + 1$$
$$1 + 1 + 1 + 1$$
$$\cdots$$

These *are* the natural numbers: the foundation upon which has been erected a beautifully complex edifice which is the theory of numbers.

The fact that one generates all the other numbers by successive additions of itself has always, even after it was no longer the only number, given one a special significance. The Greeks had a hard time defining one because it was the means by which they defined all the other numbers. Could the maker of numbers be itself a number? they asked themselves. They decided it could not. (As Aristotle reasonably observed, the measure is not measures but *the measure.*) So instead they defined one as the beginning, or principle, of number. It was so completely set apart from the other numbers that it was not considered the first odd number (that was three), but rather the great Even-Odd because when added to odd numbers it produces even and when added to even, odd. One was not a number but Number with a capital "N." It was considered to contain within itself, layer by layer like an onion, all the other numbers.†

This great reversal of *e pluribus unum* has always given one first place among the numbers in religion. During the Middle Ages, when mysticism flourished while mathematics languished, the number one represented God the Creator, the First Cause, and the Prime Mover. The other numbers were considered more imperfect in direct proportion as they receded from one. Two, as the first number so receding, signified sin which deviates from the first good. Fortunately for the larger numbers, there were ways by which they could be reduced to the digits so that they weren't completely beyond salvation.

The characteristics of the number one which give it so much nonmathematical significance are the same which make it mathematically interesting—and the same which make its behavior so obvious and hence so trivial in the ordinary processes of arithmetic.

> One is the only number which divides every number.

† The onion simile is not farfetched. Joseph T. Shipley, in his *Dictionary of Word Origins*, remarks: "Those that have, with intended humor, transposed the saying 'In *union* there is strength' to 'In *onion* there is strength' in all probability did not know that in *onion* there is *union*. With the same vowel change as in *one*, from L. *unus, one, onion* is from L. *unio, union—, unity,* from *unus.* The idea is that the many many layers make but *one* sphere. . . . The *onion* has been used as a symbol, in that, far as you may peel, you never reach the core."

One is the only number which no other number divides.

Among the infinitude of natural numbers, each in its own way unique and yet in many ways very like the others, there is no other number at all like one. The only number ever linked with it is its antithesis—zero. For while one divides all the numbers, zero divides none; while one is divided by none of the other numbers, zero is divided by all of them. Among the numbers, they are both "special cases."

The behavior of one which seems so trivial to us in multiplication and division is the direct result of one's ability to generate all the other numbers by successive additions of itself. One is the unit out of which the other numbers are built. Do not let the grammar-school obviousness of this fact deceive you, for it is the most important single fact in all the theory of numbers. When we are trying to wrest the secrets of their relationships from the numbers, the fact that one divides *all* the numbers is a most valuable weapon. It is in a sense the weapon we start with. Our next follows from it: the fact that every number is also divisible by itself.

Given the infinite set of natural numbers, each differing from its predecessor by one, the theory of numbers extends a challenge: what can be learned about the numbers besides these two easily ascertained facts that every number is divisible by one and that every number is divisible by itself?

The first step toward the understanding of any group, including the numbers, is the classification of its members into mutually exclusive subgroups. At first thought, it might not seem that the two facts we have been given would provide a basis for such a classification. It was not, indeed, the first thought of man that they did. The most ancient classification of numbers into groups was on the basis of their divisibility by the number two. The numbers that are exactly divisible by two were called *even* and those that leave a remainder of one when divided by two were called *odd*. All numbers belong in one of these groups, and no number in both. The even-odd classification seemed so basic to the Greeks that they thought of it like the great distinction between the two kinds of human beings. The even numbers they saw as "ephemeral," hence female; the odd, "indissoluble, masculine, partaking of celestial nature." But even-odd, based on divisibility of two, is not nearly so significant a classification of the numbers as that which is based on their general divisibility.

We have already made two statements about the general divisibility of numbers, and we can add to these two more which we arrive at after examining the first few numbers and their divisors.

Some numbers, like two, three, five, and seven, are divisible only by themselves and one.

Some numbers—four, six, eight, and nine, for example—are divisible also by some number other than themselves and one.

Here is a basis for a classification of numbers into two groups that has produced enough mathematics to fill most of the bulky first volume of L. E. Dickson's three-volume definitive history of the theory of numbers. The numbers in the first group, divisible only by themselves and one, are commonly called the *prime* (or first) numbers. Since it can be proved quite simply that all the numbers in the second group, divisible by some other number in addition to themselves and one, are composed of prime numbers, these are known as *composite* numbers.

(If a number n is composite, it has by definition divisors between 1 and n. If m is the least of these divisors, it must be prime because otherwise [if it is divisible by a number other than itself and one] it cannot be the least divisor of n. Continuing in this way, we can reduce all the divisors of n to primes, thus proving that every composite number can be produced by primes.)

We saw a few pages back how it is possible for us to represent all the numbers after zero by successive additions of one. Now we see that after zero and one (which, befitting their status as special cases, are neither prime nor composite ‡) we can also represent all the numbers by primes or combinations of primes.

$1 + 1$	2 (prime)
$1 + 1 + 1$	3 (prime)
$1 + 1 + 1 + 1$	2×2 (composite)
$1 + 1 + 1 + 1 + 1$	5 (prime)
$1 + 1 + 1 + 1 + 1 + 1$	2×3 (composite)
$1 + 1 + 1 + 1 + 1 + 1 + 1$	7 (prime)
. .	. . .

‡ Zero is not prime because it is divisible by an infinitude of numbers besides itself and one; not composite because, since one of its factors is always itself, it cannot be produced by primes alone. One is technically excluded from the primes because, as we shall see, if it were a prime, the most important theorem about primes would no longer be true.

We do not need to be told that the additive representation of numbers in the left-hand column is unique. It is obvious that there can be but a single possible way of expressing any number as the sum of ones. If six is $1 + 1 + 1 + 1 + 1 + 1$, it can be nothing else; and its successor among the numbers, whether we call it seven or simply the successor of six, can be nothing but $1 + 1 + 1 + 1 + 1 + 1$.

It is not, however, so obvious that the multiplicative representation of numbers in the right-hand column is also unique. Just as there is but a single way of expressing a number as the sum of ones, there is but a single way of expressing it as the product of primes.

$6 = 1 + 1 + 1 + 1 + 1 + 1$, and nothing else as the sum of ones

$6 = 2 \times 3$, and nothing else as the product of primes without respect to order.

There is but a single way that a number can be produced by primes alone. This is true of any number, no matter how large it is. A number like 17,640, for instance, is the sum of 17,640 ones and its prime factorization is $2 \times 2 \times 2 \times 3 \times 3 \times 5 \times 7 \times 7$. There are no primes except 2, 3, 5, and 7 which will divide 17,640—although, it being such a large number, we might be inclined to think that there certainly would be others. Only one combination of these four prime factors—three 2's, two 3's, one 5, and two 7's, or $2^3 3^2 5^1 7^2$—will produce 17,640. Of course, various other numbers also divide it: 6, 10, 14, 21, 35, to name a few; but ultimately these all reduce to primes—and to the primes 2, 3, 5, and 7 only.

The representation of any number as the product of primes is unique, just as the representation of any number as the sum of ones is unique.

Think for a moment of the significance of this statement. Any number can be a number so large that it has never been written out, a number so large that a man's lifetime would not be long enough to record it on paper (if the paper were long enough); any number can be any number in an infinite number of numbers. Yet, from the information we have just been given, we can make a very significant statement about this most interesting number, any number n.

We can say that n has certain prime factors, which we designate as $p_1, p_2, \ldots, p_r$, and that the prime factorization of n is a unique combination of these factors. The prime p_1 is used so many times, and we indicate this by $p_1^{k_1}$; p_2 is used so many times, and we indicate this by $p_2^{k_2}$; and so on. Just as we can say that $6 = 2 \times 3$ and $17{,}640 = 2^3 \times 3^2 \times 5 \times 7^2$, we can say of any number n that $n = p_1^{k_1} p_2^{k_2} p_3^{k_3} \ldots p_r^{k_r}$ and *know* that this representation of n as the product of primes is the only possible representation. This knowledge is so important in the study of numbers that the theorem which states it is universally acclaimed the "fundamental theorem of arithmetic."

The proof of this theorem, which tells us that the prime factorization of any n is unique, rests upon a secondary mathematical truth (known as a *lemma*) that a prime which divides the product of two or more numbers will divide at least one of those numbers. In the case of 17,640 —the number we recently used as an example—this means that 2, 3, 5, and 7, its prime factors, will each divide at least one number in any group of numbers which multiplied together produce 17,640. For example: $15 \times 28 \times 42 = 17,640$; and, in accordance with the lemma, 2 divides 28 and 42, 3 divides 15 and 42, 5 divides 15, and 7 divides 28 and 42.

The proof of the fundamental theorem itself is by *reductio ad absurdam,* a method which has been a favorite with mathematicians since the time of Euclid. It is simply assumed for the purpose of proof that prime factorization is not unique.

Let us say that a number $n = p_1^{k_1} p_2^{k_2} \ldots p_r^{k_r}$ and also $q_1^{l_1} q_2^{l_2} \ldots q_s^{l_s}$, the p's and the q's being separate sets of prime factors. On the basis of the lemma we stated above, we know that since each p divides n, which is also the product of the q's, each p must also divide some q. Since the q's are by definition prime and hence not divisible except by themselves and one, each p must be equal to some q and conversely each q equal to some p. Both sides then must contain the same primes and the prime factorization of n, contrary to our assumption, is unique.

It has been said that this theorem is essential for a systematic science of arithmetic. Certainly arithmeticians consider it so essential that for the sake of it they arbitrarily exclude the number one from the prime numbers. This is because if one is a prime, then *prime factorization of the numbers is no longer unique.* Instead of being able to say that $6 = 2 \times 3$ and nothing else as the product of primes, we would have to admit an infinite number of possible prime factorizations for six and every other number.

24

$$6 = 2 \times 3 \times 1$$
$$6 = 2 \times 3 \times 1 \times 1$$
$$6 = 2 \times 3 \times 1 \times 1 \times 1$$

. . .

Much worse would be the fact that the prime factorization of any number n could no longer be considered unique.

Because we know by the fundamental theorem that any number can be expressed uniquely in terms of its prime factors, we are able to handle n with much the same ease that we handle a particular number like six. Because we can do this, we can often prove something true about all numbers which otherwise we would have to prove for each number, one at a time, and would *never* be able to prove for *all* numbers.

The example usually given in this connection is that of the theorem which states generally which roots of which numbers are irrational. The Greeks discovered and proved that the square root of two is irrational—in other words, that it is not expressible in integers as a fraction but is a decimal which never ends and never repeats. They then went on, one at a time, to prove the irrationality of the square roots of three, five, six, seven, eight, ten, eleven, twelve, thirteen, fourteen, fifteen, and seventeen—and here stopped (the omitted numbers being squares of whole numbers). For all their labor they had proved nothing except that these few numbers out of an infinitude of numbers have irrational square roots. They had proved nothing about their other roots—cube, fourth, fifth, and so on through an infinitude of roots for each number. *With the fundamental theorem of arithmetic as a tool*, it is possible to prove simply and directly when *any* root of *any* number is irrational.§

Proving something about each number—true of one, true of two, true of three, and so on—will never, no matter how high we go, prove with finality that a statement is true about all numbers. The special challenge which the natural numbers offer man is to prove that certain things are true about all numbers without ever having a chance to examine all of the numbers individually. The extent to which man has met the challenge of the numbers rests upon the fact that one is the unit of the numbers. One sets the conditions—an infinite set with each member separated from the next by the same unit—and one puts down the weapons:

> Each number is divisible by one.
> Each number is divisible by itself.

A QUIZ

The whole subject of divisibility is basic to the study of numbers. In the course of this book we shall answer the questions below in more detail, but now the reader may enjoy trying to answer them for himself.

1. Is there a number which has no divisors?
2. How many numbers have only one divisor?
3. How many numbers have only two divisors?

§ The theorem exactly states that the mth root of N is irrational, unless N is the mth power of an integer n. In brief, we cannot possibly get a whole number by raising a fraction to any power. (No matter how often, for instance, we multiply $\frac{3}{2}$ by itself we will never get a whole number.)

4. How many numbers have an infinite number of divisors?

5. Is there a number not a divisor of any other number?

6. Is there a number which is a divisor of all numbers?

7. How many numbers divide an infinite number of numbers?

8. What is the largest number having no divisors besides itself and one?

9. How many even numbers have only two divisors?

10. After zero, what number has the greatest number of divisors?

ANSWERS

1. No. 2. Only one—one itself, since all other numbers are divisible at least by themselves and one. 3. An infinite number, since a prime has only two divisors and there is an infinite number of primes. 4. Only one—zero, which has as its divisors each and every one of the natural numbers, which are infinite. 5. Yes—zero, which can divide only itself. 6. Yes—one. 7. An infinite number, since every number except zero divides an infinite number of numbers. 8. There is no largest number, since a prime has no divisors besides itself and one and there is no infinite number of primes. 9. Only one—the number two, which is the only largest prime. 10. There is none, since we can get a number with as many divisors as we please.

27

2

THE NUMBER TWO IS NOT GENERALLY written as 10, but it can be. For two is 10 in a simple and elegant system of representing numbers known as the binary.

The binary system, which is based on two, has had something of a rags-to-riches history. It is the descendant of man's most primitive method of representing numbers as anything other than the sums of ones. It was the invention of a great mathematician who had high hopes that it might convert the Emperor of China to Christianity. Until the twentieth century it was looked upon as a mere mathematical curiosity. Now at last, as the ideal method of number representation for high-speed computing machines, it seems to have come into its own. Today it can even be seriously suggested that the ten of the decimal system should give way to some other number base, a power of two, which would simplify transposition of numbers to and from the two of the binary system.

The earliest base-two system known to man was the pair system. In the pair system there were just two number symbols, one and two. Three was two and one; four, two and two; five, two and two and one. This system was probably suggested to man by the parts of his own body. Eyes, ears, arms, legs—all were in pairs.

Although eventually he was to count by tens because he has ten fingers, he started to count by twos, perhaps because he has two hands.

The pair system, primitive though it was—and it was most primitive—met the essential requirements for a workable system of number representation. It was based on a finite number of symbols (there were only two), and it could be used to represent any number, no matter how large. It does not seem likely, though, that with the pair system man ever went beyond five.

The binary system is similar to the pair system in that it also requires only two symbols for representation of any number. The difference is that while the pair system represents numbers by twos, the binary represents them by *powers* of two.

A power of two, as of any number, is simply the result of a self-multiplication. We recall the squares and the cubes, which are second and third powers respectively, and recognize that the multiplicative process that produces them can continue indefinitely.

$$2^2 = 2 \times 2 = 4$$
$$2^3 = 2 \times 2 \times 2 = 8$$
$$2^4 = 2 \times 2 \times 2 \times 2 = 16$$
$$2^5 = 2 \times 2 \times 2 \times 2 \times 2 = 32$$

. . .

Such multiplications, even with so small a number as two, rapidly attain astronomical proportions: 2^2 is only four, but 2^{10} is in the thousands, 2^{20} in the millions, and 2^{40} (or the product of 40 twos multiplied together) in the million millions. Obviously, by using powers of two instead of twos, we can represent numbers much more compactly, and hence more efficiently. Consider the expression of even a small number like thirty. In the pair system thirty must be $2 + 2 + 2 + 2 + 2 + 2 + 2 + 2 + 2 + 2 + 2 + 2 + 2 + 2 + 2$, but in the binary it is simply $2^4 + 2^3 + 2^2 + 2^1$ $(16 + 8 + 4 + 2)$.

Except for the substitution of the powers of two for the powers of ten, the binary system works just as the decimal system does.

$$11111 \text{ (in the decimal system)}$$
$$= 10^4 + 10^3 + 10^2 + 10^1 + 10^0$$
$$11111 \text{ (in the binary system)} = 2^4 + 2^3 + 2^2 + 2^1 + 2^0$$

The difference in base gives each system an advantage over the other and a corresponding disadvantage. The decimal system, because it has the larger base, is able to represent numbers more compactly than the binary. As we see above, 11111 in the decimal system is a number 358 times as large as the decimal number 31 which is represented by 11111 in the binary system. But the binary system, because it has the smaller base, is able to represent numbers with fewer symbols. This means that it requires a smaller multiplication table to be learned, an important practical advantage.

In the binary system, the symbol 1 indicates that a particular column contains a power of two; the symbol 0 indicates that it does not. These two symbols are all that are needed, for it is possible to represent any number uniquely as the sum of powers of two. All numbers are either exactly divisible by two or divisible by two with a remainder of one. Since the zeroth power of two is one and the first power is two itself, the powers of two, it can easily be seen, are sufficient to represent any number by the use of only two symbols to indicate the presence or absence of a particular power in the column reserved for it.

(The idea that the zeroth power of two is one is hard to accept until we examine the logic behind this apparently illogical statement.

$$2^3 = 8 = 2 \times 2^2$$
$$2^2 = 4 = 2 \times 2^1$$
$$2^1 = 2 = 2 \times 2^0$$
$$2^0 = 1$$

Although we may swear we never heard of such a thing, we work every day with this concept. In the familiar decimal system, as in the binary, the first column is reserved for the zeroth power of the base—the ones.*)

The fact that *all* numbers can be represented with only 1 and 0 fascinated Gottfried Wilhelm von Leibnitz (1646–1716), the inventor of the binary system. Leibnitz was one of the greatest mathematicians who ever lived. We have only to read the account of his life in E. T. Bell's *Men of Mathematics* to be awed by the universality of his genius. Writes Bell: "The union in one mind of the highest ability in the two broad, antithetical domains of mathematical thought, the analytical and the combinatorial, or the continuous and the discrete, was without precedent before Leibnitz and without sequent after him."

All of mathematics was not enough to occupy this great mind. Leibnitz had innumerable nonmathematical projects,

* We must now be prepared to accept the idea that while every other power of zero is zero, the zeroth power of zero is one!

one of which was the reuniting of the Protestant and Catholic churches. When he invented the binary arithmetic he saw in it (according to another great mathematician, Laplace) "the image of Creation. . . . He imagined that Unity represented God, and Zero the void; that the Supreme Being drew all beings from the void, just as unity and zero express all numbers in [the binary] system of numeration." The story is that Leibnitz communicated his idea to the Jesuit who was the president of the Chinese tribunal for mathematics in the hopes that it would help convert to Christianity the Emperor of China, who was said to be very fond of the sciences.

The enthusiasm of Leibnitz for the simplicity and elegance of his system was not shared by his fellow mathematicians, for at the time it appeared that the system had nothing more than simplicity and elegance to recommend it. Yet even in Leibnitz's day the principle of representation by powers of two was commonly used by people who would never have recognized a number expressed in the binary system. These people, who knew so little of arithmetic that they did not even try to multiply except by two, had worked out their own very neat system of multiplying any numbers in this way.†

Known generally as "peasant multiplication," the system works like this. To multiply 29 by 31, divide 29 by 2 and the answer again by 2 and so on until you have only a remainder of 1. Then double 31 the same number of times that you have halved 29, keeping halvings and doublings in parallel columns. Cross out whatever doubling occurs

† Multiplication and division by two were once so commonly used that, as *duplation* and *mediation*, they were considered basic processes of arithmetic along with addition, subtraction, multiplication, and division.

opposite an even halving and add the remaining doublings to obtain your answer.

29	31
14	~~62~~
7	124
3	248
1	496
	899

If the reader will multiply 29 by 31 in the customary way, he will find that he obtains the same answer.

We can understand why the correct answer has been obtained by "peasant multiplication" only if we examine what has been done in terms of the binary system. The successive halvings of 29 have given us a binary representation of that number. All we have to do is to put a 1 after 29 itself (because it is odd), a 1 after each of the other odd halvings, and a 0 after the even halving.

29	1
14	0
7	1
3	1
1	1

We see immediately that 29 in the decimal system is 11101 in the binary. (This is the simplest method of transposing a number from decimal to binary system.)

We now see the successive doublings of 31 as multiplications by the powers of two in the binary representation of 29.

$$1 = 2^0 = 1 \qquad\qquad 1 \times 31 = 31$$
$$0 = 0 \qquad\qquad 0 \times 31 = 0$$
$$1 = 2^2 = 4 \qquad\qquad 4 \times 31 = 124$$
$$1 = 2^3 = 8 \qquad\qquad 8 \times 31 = 248$$
$$1 = 2^4 = \overline{16} \qquad\qquad 16 \times 31 = \overline{496}$$
$$\overline{29} \times 31 = \qquad\qquad 899$$

The same multiplication performed in the binary system itself looks like this.

$$
\begin{array}{r}
11111 \\
11101 \\
\hline
11111 \\
00000 \\
11111 \\
11111 \\
11111 \\
\hline
1110000011 = \\
2^9 + 2^8 + 2^7 + 2^1 + 2^0 = \\
512 + 256 + 128 + 2 + 1 = 899
\end{array}
$$

The simplicity of binary representation—the fact that in it all numbers are merely arrangements of 1's and 0's—makes it the ideal representation for high-speed computing machines.‡ Most of these so-called "giant brains" use the binary system, not because they cannot be constructed to use the decimal, but because with the decimal system their present high speed in computation could not even be approached.

‡ These are descendants of Leibnitz's calculating machine, a very superior one for its time, since it was able to do multiplication, division, and the extraction of square roots as well as addition and subtraction.

Consider a machine working with a number such as the one which for seventy-five years held the honor of being the largest known prime—a number, in spite of its great size, divisible only by itself and one. In the decimal system this number is

170,141,183,460,469,231,731,687,303,715,884,105,727.

In the binary system it is

1111111111111111111111111111111111111111111111111
1111111111111111111111111111111111111111111111111
11111111111111111111111111111111111.§

In the decimal representation, the machine must be able to differentiate between ten different possible symbols in each column of the number. In the binary representation it need differentiate only between two.

The particular usefulness of the binary representation in high-speed computation arises from the fact that the "symbols" for 1 and 0 do not have to be symbols at all. They can be simply an electric impulse for 1, to indicate the presence of a power of two in the column reserved for it, and no impulse for 0, to indicate the absence of a power.

If Leibnitz had invented his binary arithmetic especially for the computing machines of the future instead of for the Emperor of China, he could not have invented a better!

It does not matter to the machines that binary representation of large numbers takes up an astounding amount of space. It does, however, matter to the men who must transfer large numbers to and from the machines. For this task

§ This same number can be expressed more briefly in the terms of the powers of two as $2^{127} - 1$.

they have utilized still another number system in which representation is even more compact than it would be in the decimal system. This is the base-sixteen system. In it each column of number representation increases by a power of sixteen instead of by a power of two as in the binary or a power of ten as in the decimal.

$$111 \text{ (base-two)} = 2^2 + 2^1 + 2^0 = 7$$
$$111 \text{ (base-ten)} = 10^2 + 10^1 + 10^0 = 111$$
$$111 \text{ (base-sixteen)} = 16^2 + 16^1 + 16^0 = 273$$

The base-sixteen system has been selected for transposition to and from the binary because its base is a power of two (2^4). Since the binary system is based on two itself, transposition is relatively simple, as can be seen by comparing the first few powers of two in the base-two and in the base-sixteen.

2^0	(in base-two)	1	(in base-sixteen)	1
2^1		10		2
2^2		100		4
2^3		1000		8
2^4		10000		10
2^5		100000		20
2^6		1000000		40
2^7		10000000		80
2^8		100000000		100
. . .		. . .		. . .

Although in the example here, a base-sixteen representation looks exactly like a decimal representation, it does not always. For full representation in the base-sixteen we need six more symbols than the ten digits we use in the decimal system. It is common to represent ten, eleven, twelve, thirteen, fourteen, and fifteen in the base-sixteen with the last six letters of the alphabet, u standing for ten, v for eleven, and so on. The odd-looking number xyz in the base-sixteen stands for $(13 \times 256) + (14 \times 16) + (15 \times 1)$. In the base-ten the same number is represented as 3,567.

We are all so used to thinking that the decimal representation of a number "is" that number that it rarely (if ever) occurs to us that two represented as 10 is just as much two as 2. There is no special superiority in ten as a base. The superiority of modern arithmetic lies not in the ten but in the zero; positional arithmetic can be often equally efficient, and sometimes more efficient, with bases other than ten.

It has been said, in fact, that with the exception of nine, ten is probably the worst possible base for efficient number representation. (It is better than nine because while nine has only one divisor, ten has two.) A number like twelve, which has four divisors, would be a much more practical base, easily falling into halves, fourths, and *thirds*. There have been many advocates, including some royal ones, of "counting by the dozen," || but mathematicians in the past have often expressed a preference for a prime base with no divisors except the trivial ones over a base with many. With a prime base, such as seven or eleven, every fraction is expressible in only one way. Mathematicians of the present, who do a great deal of work with the high-speed computing machines and hence with the binary system, have

|| A thorough argument for twelve as a base instead of ten is presented in *New Numbers* by F. Emerson Andrews (Essential Books, New York).

another preference. Why not a number system based on a power of two?

Two itself or its square as a base would make representation, even of relatively small numbers, too lengthy. A system based on the fourth power of two, on the other hand, would involve as we have seen the addition of six new symbols. What then about a system based on the third power of two, eight?

A base-eight system does not seem like a bad idea to a great many people who work with numbers. Representation in it would be almost as compact as in the base-ten, and the multiplication table would be slightly smaller. Halves, fourths, and eighths could be easily computed. Most important, transposition to and from the binary system of the computing machines would be greatly simplified.

In spite of these arguments, it is quite unlikely that twelve, eleven, seven, or eight—or any other number—will ever replace ten as the commonly used base for number representation. But the fact that they *could* serves to remind us of something we are quite likely to forget. *A number and its symbol are not the same thing.* "Two-ness" must not of necessity be represented by 2. Whether the symbol 2 is omitted entirely in a system of number representation, as in the binary; whether a 2 in the representation of a number stands for two powers of seven, two powers of twelve, or two powers of eight instead of the usual two powers of ten; even whether some totally different symbol such as *b* is substituted for the familiar 2—the concept of the number two, the pair, remains unchanged. Two remains an interesting number.

PROBLEMS IN BINARY ARITHMETIC

It takes a little practice to perform even the simplest operations on numbers in a system other than the decimal, but there is a pleasant feeling of satisfaction in being able to do so. Below are examples of addition, subtraction, multiplication, and division as they are performed in the binary arithmetic, and then similar problems for the reader.

Addition:

$$\begin{array}{r} 100001 \\ +1011 \\ \hline 101100 \end{array} \quad \text{or} \quad \begin{array}{r} 33 \\ +11 \\ \hline 44 \end{array}$$

Subtraction:

$$\begin{array}{r} 11110 \\ -1010 \\ \hline 10100 \end{array} \quad \text{or} \quad \begin{array}{r} 30 \\ -10 \\ \hline 20 \end{array}$$

Multiplication:

$$\begin{array}{r} 1011 \\ \times 11 \\ \hline 1011 \\ 1011 \\ \hline 100001 \end{array} \quad \text{or} \quad \begin{array}{r} 11 \\ \times 3 \\ \hline 33 \end{array}$$

Division:

$$\begin{array}{r} .010101\ldots \\ 11\,\overline{)1.000000} \\ 11 \\ \hline 100 \\ 11 \\ \hline 100 \\ 11 \\ \hline 1 \end{array} \quad \text{or} \quad \begin{array}{r} .333\ldots \\ 3\,\overline{)1.000} \end{array}$$

1. In the binary arithmetic, add 110010 and 1111.
2. Subtract 11001 from 110111.

1. 1000001 (65 in the decimal system). 2. 11110 (30 in the decimal system). 3. 110010 (50 in the decimal system). 4. .00110011 . . . (.2 in the decimal system).

3. Multiply 1010 by 101.
4. Divide 1 by 101.

THREE IS AN INTERESTING NUMBER BE-
cause it is the first *typical* prime and the
primes are, as a group, the most interesting of numbers.

"It would be difficult," said a mathematician once, "for anyone to be more profoundly interested in anything than I am
in the theory of primes."

The lure of the primes has been felt by
many who are not professional mathematicians. It is perhaps a hopeful commen-
tary on human nature that men have
spent more hours thinking about prime
numbers than about assorted projects for
destroying other men. This is difficult to
understand for those who are interested
in methods of destruction.

For the primes are, after all, just numbers—numbers like two and three which
are divisible only by themselves and one.
They are the numbers from which, by
multiplication, all the other (composite)
numbers can be constructed, and for this
reason they are often called "the build-
ing blocks of the number system." With
the exception of two, they are odd, since
all even numbers after two are divisible
by the prime two and hence composite.
Three is thus, while not the first, the first
typical prime.

The distinction between these two
types of numbers, those that build and

those that are built, came relatively late in mathematics and yet it is still an ancient one. The first definition of a prime appears in the *Elements* of Euclid. Much earlier, though, it was noted that some numbers are *rectilinear* (their units being capable of arrangement only in a straight line) while others are *rectangular*.

2	3	5	7...	but	4	6	8	9...
0	0	0	0		00	000	00	000
0	0	0	0		00	000	00	000
	0	0	0				00	000
		0	0				00	
		0	0					
			0					
			0					

The rectilinear numbers, since they cannot be divided except by themselves and one, can be arranged in only one way. The rectangular numbers can be arranged in at least two ways, a straight line and a rectangle; many, like twenty-four, admit more than one rectangular arrangement.

The distinction, whether we call it prime-composite or rectilinear-rectangular, has little practical importance. For more than two thousand years it has exerted a hold on the mind of man simply because it suggests questions which are interesting but very difficult to answer.

Most of the questions are about the primes because answering a question about primes automatically answers a question about composite numbers as well.

The first question asked about prime numbers, and the first that was answered, was: How many prime numbers are there? The question, in more mathematical language, is whether the set of primes is finite or infinite. The answer has great significance for the interest of these numbers. If the primes are finite they are not nearly so interesting as if they are infinite. Theoretically, we can find out anything we want to know about a finite set of numbers by sheer physical endurance. We can even count them, no matter how many there are, because at some point there is a last one. The challenge of a finite set is merely physical. With an infinite set, the challenge is mental.

With numbers whose occurrence is regular and hence predictable, it is simple to show that they continue to appear without end. The natural numbers can be said to be infinite because we can always add one to any natural number and have another. We can always add two to an even number and have another even number, to an odd and have another odd. There is no last number, no last even number, no last odd number.

With numbers like the primes, the question of how many is much more difficult to answer. For while the natural numbers string out like beads, each the same distance from its predecessor, the same from its successor, even and odd beads alternating without exception, the prime beads occur apparently without pattern in the string of numbers.*

Even (O) and odd (X) numbers
OXOXOXOXOXOXOXOXOXOXOXOXOX...

* G. H. Hardy and E. M. Wright in their textbook *An Introduction to the Theory of Numbers* (Clarendon Press, Oxford) comment upon the fact that while the "average" distribution of the primes is very regular, their distribution in detail is extremely irregular. They suggest that for this reason the small primes would form an excellent basis for a cipher. (Another word, incidentally, that came from *sifr*, the Arabic numerals being considered at the time of their introduction into Europe a kind of secret writing.)

but

Prime (X) and composite (O) numbers
——XXOXOXOOOXOXOOOXOXOOOXOO...

The proof that the number of primes is infinite appeared in Euclid's *Elements,* almost three hundred years before the birth of Christ. It has a quality about it—a certain mathematical beauty—which even today provokes respectful envy among professional mathematicians, who cannot help asking themselves, "Would I have thought of that if it had never been thought of before?"

Euclid was an Athenian who taught for most of his life at the school in Alexandria, which he helped found. In *The Great Mathematicians* H. W. Turnbull writes, "The picture has been handed down of a genial man of learning, modest and scrupulously fair, always ready to acknowledge the original work of others, and conspicuously kind and patient." He was a man who devoted his time to numbers, not because they are useful, but because they are interesting. One of the two stories about him is that when a pupil demanded to know what he would gain by proving a theorem, Euclid ordered a slave to give him a coin "since he must make a gain out of what he learns." †

Euclid's proof that the number of primes is infinite is as straightforward as the proof that the natural numbers are infinite. It rests upon the simple fact that if we multiply together any group of prime numbers, the immediate successor of the number we get as our answer ($n + 1$, in mathematical language) will be either another prime or a

† The other story is that when the king asked him whether there was any shorter way to geometry than that set forth in the *Elements,* Euclid told his majesty, "There is no royal road to geometry!"

composite number which has as one of its factors a prime not in the group of primes we multiplied. This is because no number except one, which is not a prime, can possibly divide both n and $n + 1$.

When we multiply together any group of primes selected at random, we can observe that a new prime does indeed result; but it is most pertinent to an understanding of Euclid's proof of the infinitude of primes to observe the result when we multiply a group of consecutive primes, beginning with two and three, the first members of the set.

$2 \times 3 = 6$ and $6 + 1 = 7$, another prime
$2 \times 3 \times 5 = 30$ and $30 + 1 = 31$, another prime
$2 \times 3 \times 5 \times 7 = 210$ and $210 + 1 = 211$, another prime

Euclid's proof that the number of primes is infinite is simply this. If we take the set of what we call "all" the primes, multiply them together, and add one to our product, we will have (as above) either another prime or a composite number with a prime factor which was not included in our set. Obviously, then, we did not have "all" the primes in our set. We have now generated another, and no matter how many primes we include in the set of "all," we can always generate still another in this same way.‡

The number of primes is, therefore, infinite.

If the primes are infinite, it follows that the composite numbers are also infinite. With each additional prime we

‡ If, instead of adding one to the product of "all" the primes, we subtract one, we will get the same result.

$2 \times 3 = 6$ and $6 - 1 = 5$, another prime
$2 \times 3 \times 5 = 30$ and $30 - 1 = 29$, another prime
$2 \times 3 \times 5 \times 7 = 210$ and $210 - 1 = 209$, which has as factors primes (11 and 19) not included in our set of "all."

can build a composite number which we did not have before. We can, in fact, build an infinite number of infinite sets of composite numbers!

An example taken from the very beginning of the natural numbers will be enough to show how the composite numbers multiply with the addition of just one prime to the set of primes. Taking two as the only prime, we have as composite numbers only the powers of two.

$$4, \text{which is } 2 \times 2$$
$$8, \text{which is } 2 \times 2 \times 2$$
$$16, \text{which is } 2 \times 2 \times 2 \times 2$$
$$\cdots$$

But these powers of two are infinite in number.

With the addition of three to the set of primes, we add another set of composite numbers which are powers of three.

$$9, \text{which is } 3 \times 3$$
$$27, \text{which is } 3 \times 3 \times 3$$
$$81, \text{which is } 3 \times 3 \times 3 \times 3$$
$$\cdots$$

These too are infinite in number. By adding three to the primes, we have also added another infinite set of numbers: each of the powers of two multiplied once by three.

$$12, \text{which is } 2 \times 2 \times 3$$
$$24, \text{which is } 2 \times 2 \times 2 \times 3$$
$$48, \text{which is } 2 \times 2 \times 2 \times 2 \times 3$$
$$\cdots$$

46

In fact—and this is easily said but difficult to grasp in all its enormousness—with the addition of three, or any prime, to the set of primes we increase the set of composite numbers by an infinite number of infinite sets. Just as we multiplied each of the powers of two by three, we can also multiply each of them in turn by each of the powers of three, of which there are an infinite number.

At this point we might take a deep breath and admit that there are a lot of composite numbers.

How then does the number of primes compare with the number of composite numbers?

The primes, with two and three, start out in the lead; are even at thirteen, behind at seventeen; and continue to fall farther and farther behind. They become steadily rarer while the composite numbers become steadily more numerous. There are places in the unending series of natural numbers where we have a million, a billion, a trillion, "as many as we please" composite numbers without one prime among them. These are what are called, dramatically, the "prime deserts" and their existence can be easily proved without sending one mathematical expedition to this no man's land of number.

"As many as we please" is a favorite expression in the theory of numbers. Although it sounds a little like a grandstand gesture, it isn't. When we say that among the natural numbers there are sequences of consecutive composite numbers "as many as we please," we mean exactly that. Let us say, for simplicity's sake, that we please to find five composite numbers occurring in succession. We first multiply together the numbers from one to six (one more than the five numbers we are after) and obtain the product 720. We know then for a certainty that the five consecutive numbers 722, 723, 724, 725, and 726 are composite.§

How do we know this? We know that two divides 720, since it was one of the numbers multiplied together to produce it; if it divides 720, it must also divide 722. Therefore, 722 is composite, being divisible by at least one number other than itself and one. Since three divides 720, it must also divide 723; four must divide 724; five, 725; and six, 726. We have found "as many as we please" (which in this case happens to be five) consecutive numbers which are not prime. By exactly the same process, if we please a million instead of five, we can find a place in the sequence of natural numbers where there are at least a million composite numbers between primes. Yet there is never a number beyond which all numbers are composite.||

"Almost all" numbers are composite, but there is an infinite number of prime numbers.

Although it is often exceedingly difficult to determine whether a particular number is prime or composite, it is very easy to "make up" a number which we know in advance will be composite. We simply multiply a few primes together and there we are with a composite number. We

§ In this particular example 721 is also composite, but generally we must assume that the immediate successor of our product may be prime; for on the basis of our proof we know of no divisors for it except itself and one.

|| Although we can prove the existence of consecutive composite numbers "as many as we please," we cannot prove that there is ever a point beyond which pairs of primes separated by *only one* composite number cease to occur. The largest known pair of prime "twins" is 1,000,000,009,649 and 1,000,000,009,651. (Two and three have been called the "Siamese twins" among the primes, since they are the only two which are not separated by a composite number.)

can't do anything at all similar with primes. This is because no one has ever been able to determine a form of number which is always and invariably prime.

There have been a great many attempts to find such a generating form for primes. Not one has been successful.

How then can we tell whether a number is prime?

This is one of those deceptively simple little questions in which the theory of numbers abounds. The general method for testing a number for primality is implicit in the distinction between prime and composite numbers: if we can divide a number, it is not prime. We can test the primality of any number by the simple expedient of trying to divide it by each of the primes below the square root of the number being tested.‡ In the case of ninety-seven, this means trying two, three, five and seven. If ninety-seven is not divisible by any one of these four primes, it is not divisible by any number except itself and one.

There is a sort of assembly-line variation of this test of primality which is known as the Sieve of Eratosthenes. Eratosthenes, who lived from about 276 to about 194 B.C., is remembered particularly for an amazingly accurate measurement of the earth. His sieve was apparently the first methodical attempt to separate the prime from the composite numbers, and all subsequent tables of primes and of prime factors have been based on extensions of it. (The compilation of such a table involves a fantastic amount of work, which is not always rewarded. One table,

‡ At least one of the prime factors of a number must be equal to or smaller than its square root, since if all the prime factors were greater than the square root their product would be greater than the number itself.

published in 1776 at the expense of the Austrian imperial treasury, is reported to have had such a poor sale that the paper on which it was printed was confiscated and used in cartridges in war with Turkey.)

Using the Sieve of Eratosthenes, we can find all the primes under 100 by eliminating after two every second number; after three, every third number; and so on. This leaves us with the following numbers, which are all prime.

X	X	2	3	X	5	X	7	X	X
X	11	X	13	X	X	X	17	X	19
X	X	X	23	X	X	X	X	X	29
X	31	X	X	X	X	X	37	X	X
X	41	X	43	X	X	X	47	X	X
X	X	X	53	X	X	X	X	X	59
X	61	X	X	X	X	X	67	X	X
X	71	X	73	X	X	X	X	X	79
X	X	X	83	X	X	X	X	X	89
X	X	X	X	X	X	X	97	X	X

Besides this sieve, which facilitates finding all the primes within certain limits, and the arduous method of dividing into a particular number all possible prime divisors, there is only one completely general test for primality. The theorem which states this test bears the name, not of a great mathematician, but of a young student who subsequently gave up mathematics for law. John Wilson (1741–1793) attended Cambridge University, and it was recorded by one of his professors, Edward Waring, that he there stated what has since become known as Wilson's Theorem:

> If a number n is greater than 1, then $(n - 1)!$
> $+ 1$ is a multiple of n if and only if n is prime.

It is not thought that Wilson had proved his theorem. He had probably arrived at it by a little computation. The same theorem, it is known now, had already been stated but not published by Leibnitz. Later it was proved by several men whose names are also immortal in mathematics. However, it continued to bear the name of the young student who had enunciated it. Wilson, by the time his theorem was proved, was a judge, and if mathematics owes him a further debt it has not been recorded.

The theorem known as Wilson's is beautifully and completely general. It can be applied as a test of primality to any number, and any number which passes the test is a prime. There are more useful tests for primality than Wilson's, but none has this same quality of generality.

To test a number for primality according to Wilson's Theorem, we must first compute $(n - 1)!$. In words, such an expression means *the product of all the numbers up to and including the immediate predecessor of the number* n *which is being tested.* It is called a "factorial number," the exclamation mark being the mathematical symbol for the factorial.¶ If the prime we wish to test is 7, the product we must first obtain is $(7 - 1)!$, or 6!. This is $1 \times 2 \times 3 \times 4 \times 5 \times 6$, or 720. According to Wilson's

¶ Factorial numbers are common in formulas for permutations and combinations. The simplest of these is that $n!$ equals the numbers of different ways in which n things can be arranged. The number of different ways in which we can arrange the twenty-six letters of the alphabet, using all twenty-six every time, is 26! $(1 \times 2 \times 3 \ldots \times 26)$.

Theorem, 7 is prime if, and only if, it divides evenly $(n-1)! + 1$, or 721. Since 7 does divide 721 exactly 103 times, we know that it is prime.

The trouble with Wilson's Theorem is that it is more beautiful than useful. The great difficulty is not the size of the numbers involved, although they get very large very fast,** but the number of different operations which must be performed. It is pleasant to know that 170,141, 183,460,469,231,731,687,303,715,884,105,727 is prime if, and only if, it divides 170,141,183,460,469,231,731,687, 303,715,884,105,726 ! + 1, but even in the theory of numbers, which is not distinguished for placing a premium on usefulness, this is not considered very useful information.

The primality of this quite long number (for it is, as it happens, prime) was discovered by a completely different method. Worked out by Edouard Lucas (1842–1891) in 1876, the method, like Wilson's, tests primality without trying *any* of the possible divisors. For this reason we may discover that a number is not prime, and therefore is divisible by some number other than itself and one, and yet still not know any number which divides it.

According to Lucas, a number N of the form $2^n - 1$, where n is greater than 2, is prime if, and only if, it divides the $(n-1)$st term of a series in which the first number is 4; the second, the square of the first minus 2; the third, the square of the second minus 2; and so on—in other words, 4, 14, 194, 37634, and so on. To test the primality of 7 by this method, we must divide it into the $(n-1)$st term of the series—which, since n in the case of 7 is 3, is

** Just take the time to compute a relatively small factorial number like 26!.

52

the second number, or 14—and find it divides evenly and is therefore prime. To test the next number of this form, which is $15 = (2^4 - 1)$, we divide 15 into the third term of the series, 194, and find that it does not divide evenly and is therefore composite. The next, 31, however, divides 37634 and so is prime.

Even Lucas's method of testing primality becomes rather unwieldy when, as in the case of $2^{127} - 1$, we must divide a number like 170,141,183,460,469,231,731, 687,303,715,884,105,727 into the 126th term of the series to determine if it is prime. For numbers of such size, Lucas worked out a short cut. Instead of squaring each term of the series, he squared only the remainder after he had divided the number being tested into it. With this short cut, he was able to announce at the same time he announced his new method that he had tested $2^{127} - 1$ and found it prime.

The short cut of Lucas's method is particularly well suited to machine calculation. It was used in 1952 to test for primality the number which at the present time is the largest known prime. This number is expressed briefly as $2^{2281} - 1$. In the binary system it is represented by twenty-two hundred and eighty-one 1's. It was in this form that its primality was tested by the Bureau of Standards' Western Automatic Computer.

It used to be customary to give some idea of the size of large primes by saying that they were so many times as great as something that seemed very great in itself. But the number represented by $2^{2281} - 1$ is so large that we cannot compare it even to such a large number as that of all the electrons in the universe. The square of the electrons (a number so large that in it each electron is replaced by a universe of electrons) is equivalent to the relatively small prime $2^{521} - 1$.

There are probably few readers who will not feel a slight thrill at the thought that it is known that for all its size $2^{2281} - 1$, like three, is divisible only by itself and one. But just as some people can look at a mountain and feel no urge to climb it, not even feel vicariously another person's urge, many of us can see a large number and feel no curiosity whatsoever about whether it is prime or composite. Whatever it is that makes some people test large numbers for primality is probably somewhat like the impulse that makes a man embark upon the uncomfortable enterprise of climbing a mountain. As one famous mountain climber put it when asked why he wanted to climb a certain peak: "Because it is there."

It is fortunate for the theory of primes that there are those who are interested in testing the numbers themselves for primality. Much that is now known about primes in general was first suggested by extensive work on individual primes. Much more interesting, however, than climbing a particular mountain is finding out about mountains. Devising an efficient general test for primality is much more interesting than testing the primality of a number, no matter how large. If there is a form which invariably generates primes, that fact will be more interesting than the form itself or the primes that are generated by it. Much more interesting than the fact that $2^{2281} - 1$ is prime is the fact that there is no last prime—and this was proved at a time when any man alive would have been hard put to represent any very large number.

It was this—the theory of the primes as an infinite set,

not the individual prime numbers—to which the mathematician was referring when he said, "It would be difficult for anyone to be more profoundly interested in anything than I am in the theory of primes."

THE POWERS OF THREE

If we have three weights equal to the first three powers of three (1, 3, and 9) and if we are allowed to put a weight in either pan to balance our scale, we can weigh any number of pounds from one to thirteen inclusive. In the illustration below a □ represents the amount being weighed, and a ○, the weight:

Left			Right		
1			1		
2	1		3		
3			3		
4			3	1	
5	1	3	9		
6	3		9		
7	3		9	1	
8	1		9		
9			9		
10			9	1	
11	1		9	3	
12			9	3	
13			9	3	1

1. How many pounds can we weigh if we are allowed under the same conditions the first four powers of three as our weights?

2. How many if we are allowed the first five powers of three as weights?

ANSWERS

1. With four weights we will be able to weigh up to and including forty pounds. 2. With five weights we will be able to weigh up to and including 121 pounds.

3. With the knowledge of the number of pounds we can weigh with three, four, and five powers of three respectively, can you work out the general formula which will tell us how many pounds we can weigh when we are allowed n powers of three as our weights?

ANSWER

The answer to the general question is expressed by $\dfrac{3^n - 1}{2}$ where n is the number of powers of three used as weights.

Author's Note: Since the writing of this chapter, the number $2^{3217} - 1$ has been shown to be prime by Hans Riesel, using the computer BESK in Stockholm.

TWO TIMES TWO IS FOUR. THIS IS THE most interesting fact about the number four, and it is very interesting. Four (if we ignore the trivial 0^2 and 1^2) is the first perfect square. Four is 2^2.

There is something very solid about the symmetry of four. One of the first and most permanent number ideas was of four as the "earth number." There are still the four winds and the four elements and, of course, the four corners of the earth. Long after the world has been proved round, it carries in countless common expressions a reminder of the time when it was thought square.

The word *square* as applied to a number is a legacy from the Greeks, who looked at numbers with the eyes of geometricians. The squares to them were those numbers the units of which could be arranged in rectangles with equal sides.* These numbers with the shape of a square are, they noted, related to other numbers in several interesting ways. Each square, for instance, is the summation of successive odd

---

* Arranging the units of numbers into shapes began, according to legend, with the early Pythagoreans, who, on the sand, made people, animals, or things with pebbles and then assigned the number of the total to each representation.

numbers; and this is the way the whole series of squares can be built up, layer by layer, from a single unit. Each square is also the product of one of the natural numbers multiplied by itself.

$$0 \qquad\qquad = 1 = 1 \times 1 = 1^2$$

$$\begin{matrix} 0 & 0 \\ 0 & 0 \end{matrix} \qquad = 1 + 3 = 2 \times 2 = 2^2$$

$$\begin{matrix} 0 & 0 & 0 \\ 0 & 0 & 0 \\ 0 & 0 & 0 \end{matrix} \qquad = 1 + 3 + 5 = 3 \times 3 = 3^2$$

$$\begin{matrix} 0 & 0 & 0 & 0 \\ 0 & 0 & 0 & 0 \\ 0 & 0 & 0 & 0 \\ 0 & 0 & 0 & 0 \end{matrix} \qquad = 1 + 3 + 5 + 7 = 4 \times 4 = 4^2$$

$$\begin{matrix} 0 & 0 & 0 & 0 & 0 \\ 0 & 0 & 0 & 0 & 0 \\ 0 & 0 & 0 & 0 & 0 \\ 0 & 0 & 0 & 0 & 0 \\ 0 & 0 & 0 & 0 & 0 \end{matrix} = 1 + 3 + 5 + 7 + 9 = 5 \times 5 = 5^2$$

. . .

As a way of thinking of the squares, the representation as the second power of a number has long since replaced the geometric arrangement of the units, but the name which the eye-minded Greeks gave them has persisted.

The relationships expressed above are not the sort of thing that has kept four and the other squares interesting for more than two thousand years. Although fascinating to a people who were looking at numbers with fresh eyes, they are easily perceived and easily proved.

Difficulty in proving and difficulty in perceiving relationships between the squares and the other numbers are not, however, the only criteria for mathematical interest. Let us consider a relationship between the squares and the natural numbers that can be perceived merely by looking at them and is so obvious that it needs only the simplest proof. Yet it is a relationship that in a completely general form is the most significant in the theory of numbers.

Every number has a square. The fact does not even need a proof since it is implicit in the definition of a square as the product of a number multiplied by itself. If, then, every number has a square, the number of squares, like the number of numbers, is infinite. This was known to the Greeks. It is obviously much more easily grasped than the idea that the number of primes is infinite, an idea which they also grasped. Yet the fact that the squares, like the numbers themselves, are infinite suggested nothing more to them or to men after them until the time of Galileo Galilei (1564–1642).

Although the Encyclopedia Britannica lists Galileo as an astronomer and experimental philosopher, and this is the way we generally think of him, he was actually a professor of mathematics. The squares and the natural numbers, both unending, suggested to him a relationship which more than two hundred years after his death was to be basic to the development of the theory of the infinite. With this hint, the reader may be interested in seeing if he too will perceive the relationship implicit in the numbers below.

59

0	0^2	0
1	1^2	1
2	2^2	4
3	3^2	9
. . .	. . .	. . .

What Galileo saw first was that with the natural numbers we can *count* the squares. The zeroth square is 0; the first, 1; the second, 4; the third, 9; and so on. The disparity between the numbers with which we are counting and the squares which are being counted becomes greater as the squares become larger, the tenth square, for instance, being 100. But the important thing is that we will never run out of squares to count. There is a square for every natural number. The set of squares can be placed in one-to-one correspondence with the set of natural numbers in exactly the same way that back at the beginning of our understanding of number we placed two wolves in one-to-one correspondence with the wings of a bird.

There is a certain difference. Wolves and wings are finite and in our example there were exactly two of each. Numbers and squares are both infinite. Yet there are obviously many more numbers than there are squares, for the squares occur less and less frequently the higher we go among the numbers. We do not have to go very high to see that this is true.

$\boxed{0}$, $\boxed{1}$, 2, 3, $\boxed{4}$, 5, 6, 7, 8, $\boxed{9}$, 10, 11, 12, 13, 14, 15, $\boxed{16}$, ...

How Galileo resolved this curious contradiction is explained through the character Salviatus in *Mathematical Discourses and Demonstrations*. Having explained what we have just noted, that the squares can be placed in one-to-one correspondence with the natural numbers, a square to every number, Salviatus comes to the conclusion:

"I see no other decision that it may admit, but to say that all Numbers are infinite; Squares are infinite; and that neither is the multitude of Squares less than all Numbers, nor this greater than that; and in conclusion, that the Attributes of Equality, Majority, and Minority have no place in Infinities, but only in terminate quantities."

This conclusion of Galileo's provides modern mathematics with one of its most important definitions. On the basis of what Galileo perceived in the relationship between the squares and all the numbers we now say:

A set is called infinite when it can be placed in one-to-one correspondence with a part of itself.

This definition is just as true of the infinite set of squares as it is of the infinite set of natural numbers. If we divide the squares into even and odd, we find that we can place the members of the two subsets in one-to-one correspondence with all the squares.

Even Squares	Odd Squares	All Squares
0	1	0
4	9	1
16	25	4
36	49	9
64	81	16
. . .	. . .	. . .

We will never run out of squares, but neither will we run out of even or odd squares. We can rest assured —squares are inexhaustible.

Problems concerning squares are also inexhaustible. Even if as a group they were not, there would still be a satisfactory collection of individual problems which have been keeping mathematicians busy for a good many centuries and from all indications will continue to keep them busy. A case in point is the problem of the squares connected with what is undoubtedly the best-known theorem in mathematics:

> The square of the hypotenuse of a right triangle is equal to the sum of the squares of the other two sides.

The Pythagorean theorem, as it is usually known, was stated and proved either by Pythagoras or one of his followers some five hundred years before the birth of Christ. It was, like most of the Greek statements about numbers, geometrical. It posed, though, an interesting arithmetical problem. What are the solutions in whole numbers for the equation below?

$$a^2 + b^2 = c^2$$

One solution had been known for a long time. The Egyptians had built their pyramids by marking off a rope into 3, 4, and 5 units so that it fell automatically into the right triangle.

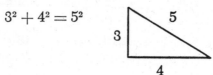

$$3^2 + 4^2 = 5^2$$

There is a way of ascertaining all possible primitive solutions to this problem. It was probably known even to the Pythagoreans, but that was by no means the end of the problem of the right triangle. Because of its relation to the squares, the Pythagorean triangle (as the right triangle with integral sides came to be known) was for centuries the basis for countless problems that, although expressed geometrically, are in reality arithmetical. Some seven centuries after Pythagoras, such problems, along with many others concerning squares and the higher powers, appeared in a little book prepared by a man known as Diophantus of Alexandria, a man whose name was to be forever linked with the squares.

Diophantus was a Greek who had an un-Greek interest in something very like algebra. Little is known about him except the problems he proposed—so little, in fact, that even the time when he lived can be estimated only in relation to the better-known lives of other men who did or did not refer to him in their own writings. His tombstone, which proposes one last problem, tells us all that is known about his personal life.

"Here you see the tomb containing the remains of Diophantus, it is remarkable: artfully it tells the measures of his life. The sixth part of his life God granted for his youth. After a twelfth more his cheeks were bearded. After an additional seventh he kindled the light of marriage, and in the fifth year he accepted a son. Elas, a dear but unfortunate lad, half of his father he was and this was also the span a cruel fate granted it, and he controlled his grief in the remaining four years of his life. By this device of numbers tell us the extent of his life."

If x is taken as the age of Diophantus at the time of

his death, the problem becomes one of solving for x in the equation

$$\frac{x}{6} + \frac{x}{12} + \frac{x}{7} + 5 + \frac{x}{2} + 4 = x.†$$

This is not the type of problem that has come to be known as a Diophantine problem. It is much too simple, there being but a single possible value for x. A more typical Diophantine problem is the ancient one of the Pythagorean triangle: to find whole number solutions for the equation $a^2 + b^2 = c^2$

The squares, particularly in connection with this very problem, were great favorites with Diophantus. One of his problems is especially interesting because it provoked, as we shall see, the theorem which has been the most difficult to prove (it has not yet been proved) in the history of mathematics. The problem appears in Book II of Diophantus' *Arithmetic* as Problem 8: *To divide a given square number into two squares.* This is the same as saying, *Given the square of the hypotenuse of a right triangle, find the squares of the other two sides*—just another of the apparently inexhaustible variations on the ancient problem.

This problem in the *Arithmetic,* and others like it, were read and struggled over for centuries before they came into the hands of the man they were all unknowingly destined for. For Diophantus of Alexandria, who died during the third century after Christ, had the honor nearly fourteen hundred years later of introducing to numbers the man who was to become known as the father of modern number theory.

† $x = 84$.

Pierre Fermat (1601–1665) was a busy, successful lawyer, thirty years old, when a copy of the *Arithmetic* of Diophantus fell into his hands. Up until that time he had apparently shown no more than a cursory interest in numbers, and he was a little old to develop a serious one. Great mathematics has almost always been produced by young, sometimes even by very young, men. We think of poets dying young, already having achieved immortality in literature. Christopher Marlowe was twenty-nine; Shelley, thirty; Keats, twenty-six. But they died no younger than some great mathematicians. Galois was twenty when he was killed in a Paris duel; Abel died in poverty in Norway at twenty-seven. Both men left behind them enough great mathematics to assure them permanent places in the history of the subject. Even mathematicians who have lived a full span must often face the fact that they did their best work when they were very young. Carl Friedrich Gauss, who was known during his lifetime, as he is known today, as the "prince of mathematicians," died at seventy-eight but he produced the *Disquisitiones Arithmeticae,* which is usually considered his masterpiece, between his eighteenth and his twenty-first years. All these mathematicians, Galois and Abel when they died, Gauss when he wrote the *Disquisitiones,* were younger than Pierre Fermat when one day he picked up the *Arithmetic* of Diophantus and got his first inkling of how very interesting the numbers are.

It has been said that Fermat was the first man to penetrate deeply into numbers. Technically never more than an amateur—his profession was the law—he is nevertheless omitted from J. L. Coolidge's *Great Amateurs in Mathematics* because, Coolidge explains, "he was so really great he should count as a professional."

For recreation the busy lawyer worked on the ancient problems of Diophantus. Usually these asked only for a single solution, but Fermat went on to state and prove theorems which gave methods for determining all possible solutions. Sometimes the problems suggested to him general theorems which stated deep and previously unsuspected relationships among the numbers.

As a mathematician, Pierre Fermat had one idiosyncrasy. Although he communicated his theorems to friends in letters or noted them down in the margins of his copy of Diophantus, he almost never stated proofs. There seems to have been no special reason that he didn't. Probably, like most mathematicians, he found what he had proved less interesting than what he was trying to prove.

In connection with Problem 8 of Book II of the *Arithmetic* he put down a characteristic note in the margin. It has been said in reference to this note that if the margin of the *Arithmetic* had been wider, the history of mathematics would have been quite different. Problem 8, as we have already stated, is *To divide a given square into two squares.* Now, Fermat was very interested in the squares, but he was also interested in the other higher powers. The problem of the squares suggested to him a much more general one, a problem that involved *all* the powers.

"On the other hand," he wrote in the margin beside Problem 8, "it is impossible to separate a cube into two cubes, or a biquadrate into two biquadrates, or generally any power except a square into two powers with the same exponent. I have discovered a truly marvelous proof of this, which, however, the margin is not large enough to contain."

(This is the same as saying that $a^n + b^n = c^n$ cannot be solved in positive integers when n is greater than two.)

Fermat's copy of the *Arithmetic* contained many other such references to proofs which were never stated. Fermat's letters to his mathematical friends were full of more. If it is curious that Fermat never offered these friends proof of the theorems which he announced with such enthusiasm, it is more than curious that they never asked for it. With anyone else but Fermat, the theorems would have probably been discounted by future mathematicians. Without a proof a theorem is not really mathematics. But Fermat was not only one of the most perceptive mathematicians who ever lived, but a mathematician of unimpeachable integrity. In every case except the one just given, when he said he had a proof for a theorem, a proof for it was later (usually much later) discovered. Only this one theorem, which is known as the Great Theorem of Fermat, remains unproved.

It is not from lack of effort. Almost all the great academies have at some time or other offered prizes for its proof. Almost every great mathematician since Fermat has tried his hand at it. Only Gauss refused, remarking tartly that he himself could state a great many theorems that nobody could prove or disprove.

Many special cases of the theorem have been proved. It has been definitely established that for prime values of n through 4001 the theorem holds. In other words, the equation $a^n + b^n = c^n$ is not solvable in positive integers when n has any prime value from three through 4001. This serves to indicate, but only to *indicate*, that Fermat was probably right, that for any n greater than two the equation is not solvable in positive integers.

Of course whether Fermat was right about the theorem is not the interesting question any more; it is whether he was right about the proof. Was he, in the seventeenth century, able to prove a theorem which, in spite of concentrated effort, no mathematician in the three centuries since has been able to prove?

At the present time it is thought that the theorem is true, but that Fermat was probably mistaken when he said he had a proof for it. (Other mathematicians since Fermat have made such mistakes.) Mathematically, it no longer matters much whether it is ever proved. It has already made its contribution, for many of the most valuable weapons of modern mathematics were forged for what invariably turned out to be unsuccessful assaults on Fermat's Great Theorem.

Pierre Fermat proved many interesting things about the squares. His famous Two Square Theorem, which is cited in any discussion of mathematical beauty, is one of the few for which he detailed his method, although even here he did not actually give his proof. The theorem states that every prime (such as five) of the form $4n + 1$ can be represented as the sum of two squares, but that no prime (such as three) of the form $4n - 1$ can ever be represented as the sum of two squares. Since all primes greater than two belong to one or the other of these forms, this is a very profound statement about the prime numbers. For proving this theorem, Fermat described his method, which he called the method of "infinite descent." He began with the assumption that there was a prime of the form $4n + 1$ which could not be represented as the sum of two squares; proved that if there was such a prime, there must then be a smaller prime which also could not; and continued in this way until he got to five, the smallest prime of the form. Since five can be represented as the sum of two squares ($1^2 + 2^2$), the assumption was obviously false; the theorem as stated, true. (Even with this assistance from Fermat himself, the Two Square Theorem was not actually proved until almost a hundred years after Fermat's death.)

The $4n + 1$ primes, incidentally, have an interesting connection with the old problem of the right triangle. Fermat also proved another theorem which states *A prime of the form* 4n + 1 *is only once the hypotenuse of a right triangle; its square is twice; its cube, three times; and so on.* As an example of this theorem, in the case of five we have

$$5^2 = 3^2 + 4^2,$$
$$25^2 = 15^2 + 20^2 \text{ and also } 7^2 + 24^2,$$
$$125^2 = 75^2 + 100^2 \text{ and also } 35^2 + 120^2 \text{ and } 44^2 + 117^2.$$

It is ironic that Pierre Fermat, who proved so many interesting things about the squares and about the other numbers, should be best known for a theorem which he quite probably did not prove. In this he reminds us of Galileo Galilei, who said so many interesting things but is best known for *Eppur si muove!* which undoubtedly he did not say.

The lives of Fermat and Galileo overlapped during the years 1601 to 1642: one man in France passing busy, relatively uneventful days as a lawyer; the other in Italy, brought before the Inquisition, tried under threat of torture, recanting on his knees his deepest scientific beliefs. They led different lives; but both, like so many other men before and after them, found the squares very interesting numbers.

AN OCCUPATION

There is nothing to keep a person occupied like trying to represent all numbers by four 4's. All four 4's must be used for every number, but various mathematical notations may also be used, as in the four examples below.

$$1 = \frac{44}{44}$$

$$2 = \frac{4 \times 4}{4 + 4}$$

$$3 = 4 - \left(\frac{4}{4}\right)^4$$

$$4 = 4 + 4 - \sqrt{4} - \sqrt{4}$$

Try now to find similar representations for five through twelve in the terms of four 4's.

ONE OF THE MOST INTERESTING THINGS about the natural numbers is that although nothing whatsoever about them changes, they retain the ability to surprise us in their relationships to one another. A case in point is that of the pentagonal numbers, those which, as their name implies, can be arranged in the shape of a five-sided figure.

The Pythagoreans had a special fondness for the shape of five. Within the regular pentagon they constructed the "triple-interwoven triangle"—the five-pointed star which was the symbol of recognition in their order. But the pentagonal numbers were to them just one group in an infinitude of so-called polygonal numbers which they found very interesting. These numbers began with three as the triangle, four as the square, and five as the pentagon and continued without end through the natural numbers. For the Greeks observed the obvious though essentially farfetched relationship that "every number from three on has as many angles as it has units."

They then observed that they could add to each first polygon a row of units and have another larger polygon of the same number of sides. Because in this row by row construction, one was the point from which construction began,

71

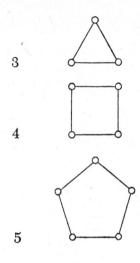

3

4

5

. . .

one was considered the first polygon in each group. In the case of five, successive pentagons were formed from a point.

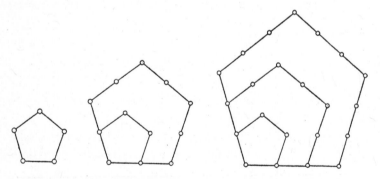

Five was thus the archetype of the pentagonal numbers, having as many units as a pentagon has angles; but one was the first. The pentagonal numbers immediately following one and five appear below.

$$12, 22, 35, 51, 70, 92, 117, 145, 176, 210, \ldots$$

Since a standard test of mathematical aptitude is the ability to see the basis for selection in a series of numbers like the one above, the reader might like to try to continue the series by adding the appropriate next number to it. The next number, as it happens, is the thirteenth pentagonal number. There is more than one way of arriving at its value.

The first is by summing one and every third number up to and including the thirteenth. As we saw in "Four," the squares are successive summations of every second number after one. The pentagonals, we now see, are the successive summation of every third (the hexagonals, of every fourth; and so on).

Pentagonal
$$\# 1 = 1 = 1$$
$$\# 2 = 1 + 4 = 5$$
$$\# 3 = 1 + 4 + 7 = 12$$
$$\# 4 = 1 + 4 + 7 + 10 = 22$$
$$\ldots$$
$$\# 12 = 1 + 4 + 7 + 10 + 13 + 16 + 19 +$$
$$22 + 25 + 28 + 31 + 34 = 210$$

To obtain the thirteenth pentagonal number (247) we add 37 (34 + 3) to the twelfth pentagonal number above.

The second way of arriving at the thirteenth pentagonal number is more direct, but involves a knowledge of the general formula for determining any polygonal number. In the language of mathematics, any polygonal number is referred to as the nth r-agonal number. In our case we are after the 13th 5-agonal number. In the formula below then the letter n has the value of 13, the letter r of 5.

$$P_n^r = \frac{n}{2}[2 + (n-1)(r-2)] \text{ or } n + (r-2)n\frac{(n-1)}{2}$$

or

$$\text{13th 5-agonal number} = \frac{13}{2} \times 38 \text{ or } 13 + (3 \times 78) = 247$$

Like many after them who have looked at numbers for the first time, the Greeks found the formation and inter-relationships of the polygonal numbers very interesting. More sophisticated mathematicians, once they have the above formula with which they can obtain any polygonal number of any rank, are inclined to dismiss them all as the kind of thing that only amateurs would find interesting. But no one can dismiss as uninteresting numbers which Pierre Fermat found interesting. Fermat looked at the natural numbers with something of the fresh interest of the Greeks, but went below the surface relationships of the polygonal numbers that had intrigued them and discovered a relationship between the polygonals and all the numbers which they had never dreamed existed.

"Every number," wrote Pierre Fermat in the margin of his copy of Diophantus, "is either triangular or the sum of two or three triangular numbers; square or the sum of two, three, or four squares; pentagonal or the sum of two, three, four, or five pentagonal numbers; and so on."

The beauty of this theorem lies in the *every number* and the *and so on*. It is completely general. It says something about all numbers and about all polygonal numbers, and what it says is not at all obvious. J. V. Uspensky and M. A. Heaslet, who in their *Elementary Number Theory* dismiss the Greek interest in the polygonal numbers as trivial, say of Fermat's theorem, respectfully, "This is a truly deep property of numbers."

That every number can be expressed as the sum of five or fewer pentagonal numbers is an unexpected relationship between these numbers of the shape of five and all the natural numbers. It lacks, however, that quality of surprise which we find in an even later discovery.

The discoverer was no amateur in mathematics, even in the grand sense that Pierre Fermat was. Leonhard Euler (1707–1783) was one of the most completely professional mathematicians who ever lived and, without a rival, the most prolific.* During his seventy-six years there was scarcely an aspect of mathematics which he did not leave more systematized than he had found it, this in spite of the fact that the last seventeen years of his life were spent in almost total blindness.

It was one of the rare things about Euler that he did what needed to be done—whether it was making a great original contribution or merely picking up pins to keep things straight in mathematics. One thing that was needed at the time of Euler—and the thing we are especially interested in —arose out of the study of partitions. It was there, where no one would have expected to find the pentagonal numbers playing an important part, that Euler found them.

In the theory of partitions, we are concerned with the number of ways in which a number can be represented as the sum of its parts. The partitions can be restricted to particular parts (such as odd parts or distinct parts), but most generally they are without restriction. How many

* The project of publishing Euler's collected mathematical papers is still uncompleted. Calculations of the cost were badly upset by the discovery, more than a century after Euler's death, of an unsuspected store of manuscripts.

different ways, for instance, can the number five be represented as the sum of the numbers one, two, three, four, and five?

$$5$$
$$4 + 1$$
$$3 + 2$$
$$3 + 1 + 1$$
$$2 + 2 + 1$$
$$2 + 1 + 1 + 1$$
$$1 + 1 + 1 + 1 + 1$$

The number of unrestricted partitions for five is seven, or $p(5) = 7$.

The general problem in the theory of partitions is to determine the number of partitions possible for each of the natural numbers. This is no simple task, for the number of partitions bears no fixed relationship to the number being partitioned. One can be partitioned in only one way, two in two ways, three in three ways; but after three the relationship is no longer one-to-one. The number of partitions for four is five; for five, as we have seen, seven. The reader may be interested in hazarding a guess as to the number of ways in which six can be partitioned. Unless he actually computes them, he is almost certain to come up with the wrong answer.†

If there is no apparent relationship between a number and the number of its partitions, how then can we determine the partitions of a particular number without actually computing them? What would be useful would be a combination of numbers which by multiplication or division, or

† $p(6) = 11$.

both, would automatically produce the answers we are after. *Producing answers indefinitely.* For we want the number of partitions for each and every one of an infinitude of numbers!

It was just such a "generating function" that Leonhard Euler contributed to the theory of partitions. In discovering it, he discovered also a very surprising relationship between the pentagonal numbers and the unrestricted partitions of all the natural numbers.

Euler's generating function for $p(n)$ is the reciprocal of a power series.

$$\frac{1}{(1 - x)(1 - x^2)(1 - x^3)(1 - x^4)(1 - x^5)\ldots}$$

The expression indicates, as does any fractional expression, that a division is to be performed. There will, however, be something very odd about this division. We will have to start our division without ever having completed our divisor! We are instructed by the three dots after $(1 - x^5)$ that we are to continue multiplying in the same fashion. Each time we are to increase the power of x by one so that we will be multiplying $(1 - x^5)$ by $(1 - x^6)$ by $(1 - x^7)$ and so on and on. The terms being multiplied will never end and neither will the product which we obtain with them. More and more of it will be finally established the longer we multiply, but all of it—never. It must be what is known as an infinite product.

When therefore we divide this infinite product into one, we will obtain as our answer an infinite quotient. This is as it should be, for the generating function of $p(n)$ must by definition never stop generating. It has to produce for us the number of partitions for every one of the endless string of natural numbers.

All in all, this generating function for $p(n)$ is an odd sort of thing to anyone used to multiplication and division with finite numbers. The multiplication which gives us the infinite product looks odd enough, but the division which gives us the infinite quotient looks odder yet.

We begin, with deceptive ease, by multiplying $(1 - x)$ by $(1 - x^2)$. The early part of the multiplication is printed in full below so that the reader can see how in a short time the first few terms of our answer no longer change with continued multiplication. They are, in other words, established; and we can use them to begin our division of one.

$$1 - x$$
$$1 - x^2$$
$$\overline{1 - x - x^2 + x^3}$$
$$1 - x^3$$
$$\overline{1 - x - x^2 + x^3}$$
$$\qquad\qquad\quad - x^3 + x^4 + x^5 - x^6$$
$$\overline{1 - x - x^2 \qquad + x^4 + x^5 - x^6}$$
$$1 - x^4$$
$$\overline{1 - x - x^2 \qquad + x^4 + x^5 - x^6}$$
$$\qquad\qquad\qquad\quad - x^4 + x^5 + x^6 \qquad\quad - x^8 - x^9 + x^{10}$$
$$\overline{1 - x - x^2 \qquad\qquad + 2x^5 \qquad\qquad - x^8 - x^9 + x^{10}}$$
$$1 - x^5$$
$$\overline{1 - x - x^2 \qquad\qquad + 2x^5 \qquad\qquad - x^8 - x^9 + x^{10}}$$
$$\qquad\qquad\qquad\qquad\quad - x^5 + x^6 + x^7 \qquad\qquad - 2x^{10} \ldots$$
$$\overline{1 - x - x^2 \qquad\qquad + x^5 + x^6 + x^7 - x^8 - x^9 - x^{10} \ \ldots}$$
$$1 - x^6$$
$$\overline{1 - x - x^2 \qquad\qquad + x^5 + x^6 + x^7 - x^8 - x^9 - x^{10} \ldots}$$
$$\qquad\qquad\qquad\qquad\qquad\quad - x^6 + x^7 + x^8 \qquad\qquad - x^{11} - x^{12}$$
$$\overline{1 - x - x^2 \qquad\qquad + x^5 \qquad + 2x^7 \qquad - x^9 - x^{10} - x^{11} - x^{12} \ldots}$$
$$1 - x^7$$
$$\overline{1 - x - x^2 \qquad\qquad + x^5 \qquad + 2x^7 \qquad - x^9 - x^{10} - x^{11} - x^{12} \ldots}$$
$$\qquad\qquad\qquad\qquad\qquad\qquad\qquad - x^7 + x^8 + x^9 \qquad\qquad\qquad - x^{12} \ldots$$
$$\overline{1 - x - x^2 \qquad\qquad + x^5 \qquad + x^7 + x^8 \qquad - x^{10} - x^{11} - 2x^{12} \ldots}$$
$$1 - x^8$$
$$\overline{\qquad\qquad\qquad\qquad\qquad\qquad\qquad\qquad\qquad\qquad\qquad\qquad}$$
$$\ldots$$

The first of these established terms is:

$$1 - x^1 - x^2 + x^5 + x^7 - x^{12} - x^{15} + x^{22} + x^{26} \ldots$$

By substituting a 1 for each of these powers of x which has remained in the product and substituting a 0 for each which has dropped out, we get an odd-looking representation for the beginning of our divisor.

$$1 - 1 - 1 + 0 + 0 + 1 + 0 + 1 + 0 +$$
$$0 + 0 + 0 - 1 + 0 \ldots$$

We are now ready to divide unity.

$$
\begin{array}{r}
1 + 1 + 2 + 3 + 5 + 7 + 11 \ldots \\
\hline
1 - 1 - 1 + 0 + 0 + 1 + 0 \ldots \,\big|\, 1 + 0 + 0 + 0 + 0 + 0 + \ 0 \ldots \\
1 - 1 - 1 + 0 + 0 + 1 + \ 0 \ldots \\
\hline
+ 1 + 1 + 0 + 0 - 1 + \ 0 \ldots \\
1 - 1 - 1 + 0 + 0 + \ 1 \ldots \\
\hline
+ 2 + 1 + 0 - 1 - \ 1 \ldots \\
2 - 2 - 2 + 0 + \ 0 \ldots \\
\hline
+ 3 + 2 - 1 - \ 1 \ldots \\
3 - 3 - 3 + \ 0 \ldots \\
\hline
+ 5 + 2 - \ 1 \ldots \\
5 - 5 - \ 5 \ldots \\
\hline
+ 7 + \ 4 \ldots \\
7 - \ 7 \ldots \\
\hline
+ 11 \ldots
\end{array}
$$

From the fragment of the division which we have reproduced the reader will observe that the numbers appearing in our answer seem strangely familiar. They are familiar, for they are in order the number of unrestricted partitions possible for each of the first few natural numbers.

$$p(0) = 1$$
$$p(1) = 1$$
$$p(2) = 2$$
$$p(3) = 3$$
$$p(4) = 5$$
$$p(5) = 7$$
$$p(6) = 11$$
$$\cdots$$

Continuing the division of one in this manner will continue to produce the successive values of $p(n)$. It will even turn up casually such a fact as that the number of unrestricted partitions for a relatively small number like 200 is 3,972,999,029,388.

In arithmetic like this we have no reason to expect to find our old friends, those trivial, if amusing, numbers which can be arranged into the shape of five. Yet if we will examine the first few terms of our infinite product which have been definitely established, we will find

$$1 - x^1 - x^2 + x^5 + x^7 - x^{12} - x^{15} + x^{22} + x^{26} \cdots$$

Now let us do a little arithmetic of our own. The formula for a pentagonal number is

$$P^5_{\ n} = \frac{3n^2 - n}{2}.$$

Up until now we have been considering only the pentagonal numbers produced by this formula when the value of n is 0 or one of the positive integers.

1 when $n = +1$
5 when $n = +2$
12 when $n = +3$
22 when $n = +4$

. . .

But the same formula also yields pentagonal numbers for negative values of n.

2 when $n = -1$
7 when $n = -2$
15 when $n = -3$
26 when $n = -4$

. . .

If we now re-examine the first few established terms of our infinite product, we will find that *the only x's which remain are those the exponents of which are the pentagonal numbers produced by the formula above for both the negative and positive values of* n.

This is not a relationship that Euler or anyone else suspected. It is curious and not exactly clear why the pentagonal numbers should make this appearance in the generating function for $p(n)$. But it is a discovery that the Greeks would have appreciated. For if they were distracted sometimes by the more trivial and obvious aspects of number composition, they were nevertheless the first people to realize that the numbers as numbers are fascinatingly complex in their relationships—and full of surprises.

It is very satisfying to discover for oneself interesting relationships among the numbers even though the same relationships have already been discovered by somebody else. If the reader will square the beginning of the infinite product which we obtained in this chapter, he will find nothing interesting about the result; but if he will cube it, he will discover a surprising and interesting pattern. The first to discover it was C. G. J. Jacobi (1804–1851), a very great mathematician.

$$1 - x - x^2 + x^5 + x^7 - x^{12} - x^{15}$$
multiplied by
$$1 - x - x^2 + x^5 + x^7 - x^{12} - x^{15}$$
multiplied by
$$1 - x - x^2 + x^5 + x^7 - x^{12} - x^{15}$$
yields
?

ANSWER

$1 - 3x + 5x^3 - 7x^6 + 9x^{10} \cdots$, the exponents which remain in this case being the triangular numbers rather than the pentagonal numbers and the coefficients being alternately the positive and negative odd numbers.

Six is the first "perfect" number.
The Greeks called it perfect because it is the sum of all its divisors except itself. These are one, two, and three; and $6 = 1 + 2 + 3$.*

The Romans attributed the number six to the goddess of love, for it is made by the union of the sexes: from three, which is masculine since it is odd, and from two, which is feminine since it is even. The ancient Hebrews explained that God chose to create the world in six days instead of one because six is the more perfect number.

Perfect numbers have been interesting to mathematicians, and to others, since the time of the Greeks. But starting out with six, mathematicians in more than two thousand years turned up only eleven more which met the strict requirements for numerical perfection. Then in 1952 a University of California professor, working with the National Bureau of Standards' Western Automatic Computer at what was then the Institute for Numerical Analysis on the Los Angeles campus of the uni-

---

* The Pythagoreans hailed ten as "perfect." It is not perfect in the way that six is, but it had for them the special charm of containing all the geometric forms. It is the sum of one (the point), two (the line), three (the plane), and four (the solid).

versity, discovered the first new perfect number in seventy-five years and, in the next few months, four more for a total of seventeen.

The discovery did not attract the attention of the press. Perfect numbers are not useful in the construction of bombs. In fact, perfect numbers are not useful at all. They are merely interesting, and their story is an interesting one.

It begins, like most stories in mathematics, with the Greeks, who, having noticed the fact that six $(1 + 2 + 3)$ and twenty-eight $(1 + 2 + 4 + 7 + 14)$ are both the sum of all their divisors except themselves, wondered how many other numbers there are like them. The basic similarity of six and twenty-eight is apparent when both are represented algebraically. They are of the form $2^{n-1}(2^n - 1)$.

$$6 = 2^1(2^2 - 1), \text{ or } 2 \times 3$$
$$28 = 2^2(2^3 - 1), \text{ or } 4 \times 7$$

Euclid, more than two thousand years ago, proved that all numbers of this form are perfect when $2^n - 1$ is divisible only by itself and one or, in other words, prime. In the case of six, we see from the above that the prime essential to its formation is $3 = (2^2 - 1)$; in the case of twenty-eight, $7 = (2^3 - 1)$. Euclid did not, however, prove that *all* perfect numbers are of this form, and he left for future mathematicians a question:

How many perfect numbers are there?

In the centuries that followed, the numbers seemed to gain more ethical than mathematical significance. L. E. Dickson in his history of the theory of numbers reports that in the first century A.D. numbers were separated into *abundant* (those like twelve whose divisors total more than themselves), *deficient* (those like eight whose divisors total less), and *perfect;* and the moral implications of the three types were carefully analyzed. In the eighth century it was pointed out that the second origin of the human race was made from the deficient number eight, since in Noah's Ark there were eight human animals from whom the entire human race sprung, this second origin being thus more imperfect than the first, which was made according to the perfect number six. In the twelfth century the study of perfect numbers was recommended in a program for the "Healing of Souls."

Nobody, however, answered Euclid's question.

Nobody, in fact, seemed much concerned with the mathematics of the subject. The first four perfect numbers—6, 28, 496, and 8,128—had been known as early as the first century. The basic theorem concerning perfect numbers had been carefully enunciated by Euclid three hundred years before the birth of Christ. Yet it was not until fourteen centuries later, in spite of all the speculation on the subject in general, that the fifth perfect number was correctly stated for the first time as 33,550, 336.

Looking smugly back from an age of high-speed computing machines, we may forget that the discovery of perfect numbers, even those much smaller than the largest ones known today, has always involved a considerable amount of computation. Let's take this fifth perfect number as a case in point. From it the reader can get some idea of just how much of the third "R" was behind the announcement of it as perfect.

To prove that $2^{13-1}(2^{13}-1)$, the representation of 33,550,336 according to Euclid's formula, is perfect, we must prove that $2^{13}-1$ is prime. First we must compute the number represented by $2^{13}-1$—a multiplication of thirteen 2's, which gives us the product 8,192, which, less 1, gives us $2^{13}-1$ as 8,191.

To ascertain whether 8,191 is prime, we must try to divide it by all the primes below its square root, which falls between 90 and 91. There are twenty-four primes below 90. Only after we have verified that none of these divides 8,191 can we say that it is prime. For this job we need an accurate listing of the primes and accurate reckoning at every step. When we recall that reckoning up until this time was done without a practical system of arithmetic notation, we do not marvel that it was so long until the fifth perfect number was accurately stated. For *after* we have ascertained that 8,191 is indeed prime, we must multiply it by 4,096 (or 2^{12}) to obtain 33,550,336.

The reader with time to spare may be interested in trying his skill on $2^{17-1}(2^{17}-1)$, which is the next possible perfect number of Euclid's form.

Because perfect numbers are, after the fourth, so large and offer so many opportunities for error in computation, a great many imperfect numbers were announced at various times as perfect.

There was a tendency also to guess at the unknown numbers from the known. On the basis of the first four perfect numbers—6, 28, 496, and 8,128—two guesses were widely accepted. One was that the perfect numbers ended alternately in 6 and 8. As it happens, they do end in 6 or 8, but not alternately or in any discernible pattern. This hypothesis went by the board with the announcement of the sixth perfect number—8,589,869,056—which ends in 6 when it "should" end in 8. The other guess was that they appeared in regular fashion throughout the numbers, one (6) in the units, one (28) in the tens, one (496) in the hundreds; one (8,128) in the thousands. The discovery of the fifth perfect number, which left the ten thousands, the hundred thousands, and the millions without their allotted numbers, disproved this hypothesis.

In the perfect number business, though, anybody's guess appears to have been as good as anybody else's. The fact that a guess was wrong did not write it off the records. The particular primes of the form $2^n - 1$ which are necessary for the formation of even perfect numbers bear for all time the name of a man who guessed wrong.

Marin Mersenne (1588–1648) was a friar whose best claim to mathematical importance lay in the fact that he was a favorite correspondent of both Fermat and Descartes. It was in 1644 that he established another claim and linked his name forever with the perfect numbers. By that time it had been proved that for $2^n - 1$ to be prime, n must be prime. Since with the fifth perfect number even the necessary primes were enormous, it had become necessary to describe all perfect numbers by the prime values of n in the expression $2^n - 1$. The five known perfect numbers were by this system designated as 2 (for the exponent in $2^2 - 1$, the prime necessary for the formation of 6), 3, 5, 7, and 13. Mersenne now announced that there were only six more such prime exponents up to and including 257. He listed them as 17, 19, 31, 67, 127, and 257. This last and largest number he announced as prime ($2^{257} - 1$) is

231584,178474,632390,847141,970017,375815,706539,
969331,281128,078915,168015,826259,279871.

It was obvious to other mathematicians that Father Mersenne could not have tested for primality all the numbers which he had announced as prime. But neither could they. One contemporary suggested hopefully that the basis of Mersenne's assertion was doubtless to be found in his stupendous genius which perhaps recognized more truths than he could demonstrate.

At the time Mersenne announced his "primes" the only method of testing the primality of a number was the one previously mentioned of actually dividing into it all the primes smaller than its square root. This was a method so time consuming that for some of Mersenne's numbers even the high-speed computing machines of the distant future would not be able to achieve results with it. But by this laborious method mathematicians did test the primes announced by Mersenne for the sixth, seventh, and eighth perfect numbers. The eighth ($2^{31} - 1$) was tested and found prime by Leonhard Euler, busy as usual doing what needed to be done in mathematics.

A mathematical writer commented that the perfect number formed from Euler's prime would quite probably be the last to be discovered. "For, as [the perfect numbers] are merely curious without being useful, it is not likely that any person will attempt to find one beyond it." Little did he reckon with the curiosity of mathematicians when the question is whether a particular kind of number is finite or infinite!

It was Euler who made the most important contribution since Euclid to the question of the perfect numbers.

Euclid had proved that any number of the form 2^{n-1} $(2^n - 1)$ is perfect when $2^n - 1$ is prime, but he did not prove that all perfect numbers are of this same form. Euler proved that all *even* perfect numbers are. There are, as far as is known, no odd perfect numbers; but it has never been *proved* that there are none.† (There are none, it is known, below 2,000,000.)

Euler's perfect number remained the largest known for more than a hundred years. Then in 1876 Edouard Lucas worked out the method which we have already described in "Three" by which a possible prime can be tested without factoring. At the same time that he announced his method, he announced that he had tested $2^{127} - 1$ and found it prime. Although in 1891 in his *Théorie des Nombres* he changed his mind and listed the number as "undecided," after verification in 1913 it was accepted until the discoveries in 1952 as the largest known Mersenne prime. (Until 1951, it was also the largest known prime of any type.)

Even with the help of Lucas's much more efficient method of testing Mersenne numbers, mathematicians were not able to finish testing all of them until a few years ago. The last, $2^{257} - 1$ (printed in full on page 88), required over a year of work on a standard calculating machine and then another year for checking the result. It is not prime. Since it was the largest guess Mersenne made, the final score on the mathematician-friar could at last be reckoned. In addition to the five perfect numbers already known at the time he made his famous announcement, he had listed four more cor-

---

† If there is an odd perfect number, it must have no less than six distinct prime factors.

rectly (17, 19, 31, and 127), two incorrectly (67 and 257), and had omitted three which he should have included since they are below 257 (61, 89, and 107).

As the twentieth century went into its second half, there were twelve known perfect numbers, the largest being $2^{126}(2^{127} - 1)$, which had been discovered seventy-five years before by Lucas. There had been one venture beyond Mersenne's upper limit of 257 by high-speed computing machine, but it had yielded no new numbers. And Euclid's question was still unanswered.

The machine which in 1952 broke the barrier in the Mersenne numbers was the National Bureau of Standards' Western Automatic Computer, known briefly as the SWAC. It is one of the fastest computing machines in existence. It can add two billion-dollar figures in sixty-four microseconds, a microsecond being one millionth of a second. This means that the SWAC can do an addition of this type 156,000 times faster than a man can—if a man can do it in ten seconds.

The mind work necessary to prepare the problem of the Mersenne primes for this high-speed machine was done by R. M. Robinson, a professor of mathematics at the University of California. Getting the problem ready for the machine took several months.

Contrary to the popular picture, a so-called giant brain can't *think* any more than a so-called giant eye can see. The SWAC extends man's ability to compute just as the Palomar telescope (which is its neighbor in southern California) extends his ability to see. That is all. The SWAC is no mathematician. Except for its phenomenal speed and accuracy, it is inferior to any human being who knows how to add, subtract, divide, and multiply effi-

90

ciently. For it cannot compute anything it has not been told how to compute.

Professor Robinson's job was to break down the Lucas method of testing primality into a program of the thirteen kinds of commands to which the SWAC responds. The job was complicated by the fact that while the machine was built to handle numbers of thirty-six binary digits, the problem numbers ran to twenty-three hundred such digits. It was, Professor Robinson found, very much like trying to explain to a human being how to multiply hundred-digit numbers on a desk calculator built to handle ten-digit numbers. One hundred and eighty-four separate commands were necessary to tell the SWAC how to test a possible prime by the Lucas method. The same program of commands, however, could be used for testing any number of the Mersenne type from $2^3 - 1$ to $2^{2297} - 1$, the largest which could be handled on the machine.

There was still more for man to do before the machine could "solve" the problem. The commands had to be coded. This was done by using the letters and signs of the standard typewriter keyboard, the letter "a," for example, being the code letter for the command to add. Coded, the commands were then transferred to a heavy paper tape so that they became merely an arrangement of perforations which could be recognized by the machine either as an electric impulse (a hole punched in the tape) or as the lack of an impulse (no hole).

Such simplicity of language is the main factor in the SWAC's amazing computing speed. Even the enormous numbers it works with are expressed wholly as 1's (impulses) or 0's (no impulses). The SWAC, instead of us-

91

ing the decimal system for its computations, uses the binary system which we described in "Two."

On the evening of January 30, 1952, the program of commands, coded and punched on a twenty-four-foot tape, was placed in the machine. In comparison to the seconds it would take the SWAC to obey all the commands on the tape, the insertion itself took an extremely long time—several minutes. All that was now necessary to test the primality of any Mersenne number was to insert the exponent of the new number as it was to be tested. The machine could do the rest, even to typing out the result—continuous zeros if the number was a prime, a number written out to the base-sixteen if it was not. The proof of primality would be in the string of continuous zeros because by the Lucas test (described in "Three") a number is prime only if it leaves no remainder when divided into a certain term in a certain series.

The human operator of the SWAC, sitting at a desk in front of the large machine, inserted the first number to be tested. He typed it out backwards, not in the binary system, which would have made it too lengthy, but in the base-sixteen so that the machine itself could transpose it into the binary. He then pressed a button on the panel of his desk, and the machine, following the one hundred and eighty-four instructions it had received, began the test for primality of the first number.

The first number selected was $2^{257} - 1$, the largest of the eleven numbers announced as prime by Mersenne. Twenty years before the SWAC test, it had been tested and found not prime by D. H. Lehmer. It had taken him two hours a day for a year to make the test. This evening it happened that Lehmer himself, now director of research at the Institute of Numerical Analysis, was in the room to see the machine, in a fraction of a second, arrive at the answer that had taken him an arduous seven hundred and some hours: $2^{257} - 1$ is not prime.

The SWAC then continued on a list of larger possible primes. Mersenne had said, four hundred years before, that to tell if a given number of fifteen or twenty digits is prime, all time would not suffice for the test; but he had not foreseen a short cut like Lucas's or a machine like the SWAC. One by one, by Lucas's method, it tested forty-two numbers, the smallest having more than eighty digits! Not one proved prime.

It was not until ten o'clock that evening that the long awaited string of zeros came out of the machine—135 of them. The number just tested, briefly expressed as $2^{521} - 1$, was the first Mersenne prime in seventy-five years. The new perfect number which could be formed from it—$2^{520}(2^{521} - 1)$—was only the thirteenth to be discovered in almost twice that many centuries.

For a period of approximately two hours on the night of January 30, 1952, $2^{521} - 1$ had the distinction of being as well the largest known prime number. Then shortly before midnight the string of zeros announcing another, larger prime came up again. In the next few months the SWAC, testing to its upper limit, turned up three more —for a total, which stands at the present time, of seventeen perfect numbers. The testing of the new thirteenth Mersenne prime had taken the machine approximately one minute—the equivalent of a year's full-time work for a man. The seventeenth and last took the machine an hour. It would have taken a man a lifetime.

The seventeen perfect numbers:

$$2(2^2 - 1)$$
$$2^2(2^3 - 1)$$
$$2^4(2^5 - 1)$$
$$2^6(2^7 - 1)$$
$$2^{12}(2^{13} - 1)$$
$$2^{16}(2^{17} - 1)$$
$$2^{18}(2^{19} - 1)$$
$$2^{30}(2^{31} - 1)$$
$$2^{60}(2^{61} - 1)$$
$$2^{88}(2^{89} - 1)$$
$$2^{106}(2^{107} - 1)$$
$$2^{126}(2^{127} - 1)$$
$$2^{520}(2^{521} - 1)$$
$$2^{606}(2^{607} - 1)$$
$$2^{1278}(2^{1279} - 1)$$
$$2^{2202}(2^{2203} - 1)$$
$$2^{2280}(2^{2281} - 1)$$

The smallest untested Mersenne number now is $2^{2309} - 1$. It is not, however, so interesting as the Mersenne number $2^{8191} - 1$. The exponent here is itself the Mersenne number $2^{13} - 1$ (8191); and—according to a conjecture which has proved true in the first four cases—whenever a Mersenne number has as its exponent a Mersenne number, it is prime.‡ A recent test made of this number on another machine called the Illiac, at the University of Illinois, indicated that it is composite and the conjecture, interesting as it is, false. This test took over one hundred hours.

The largest known perfect number, $2^{2280}(2^{2281} - 1)$, formed from the seventeenth Mersenne prime discovered

‡ For the exponents 3, which is $2^2 - 1$; 7, which is $2^3 - 1$; 31, which is $2^5 - 1$; and 127, which is $2^7 - 1$.

with the SWAC in 1952, has been computed in the decimal system by Horace S. Uhler as:

```
  994 97054 33708 64734 42435 20260 45228 16989 64386
35711 26408 51177 40205 75773 84932 63555 29178 68662
94981 51336 41650 25166 45641 69951 68131 40394 89794
06365 61646 54594 77532 32301 45360 35832 23268 08561
36472 33768 08164 57276 69037 39438 56965 22820 30153
58880 41815 55951 34080 36145 12387 05843 25525 81395
04871 09647 77074 38273 62571 82287 05676 43040 18472
31158 25645 59038 63133 77067 11263 81492 53171 84391
47800 65137 37344 62224 06322 95356 91247 71480 10136
31809 66448 09988 22924 53452 39542 82708 75732 53631
15392 66115 11649 07049 40164 19241 77449 19250 00089
47274 07937 22982 93005 78253 42788 44943 58459 94953
52318 19781 36144 96497 79252 94809 99098 21642 20748
55148 05768 28811 55834 09148 96987 57905 23961 87875
31249 72681 17994 42346 41016 96001 18157 88847 43661
01927 04551 63703 44725 52319 82033 65320 14561 41202
88204 92176 94041 83770 74274 38914 99243 03484 94544
61051 21267 53806 15832 99291 70797 23788 07395 01603
07654 40655 60175 91093 70564 52264 79891 56121 80427
30122 66011 78345 11022 30081 38040 19513 83582 98714
95782 29940 81818 15140 46314 81931 32063 21375 97333
67850 23565 44310 13056 33127 61023 05495 88655 60595
13323 51485 64175 75426 11227 10807 32633 89434 40959
59768 35137 41218 70253 49639 50440 40616 54653 75534
91626 80629 29055 16441 53382 76068 18622 94677 41498
90474 91922 79570 72109 20437 81113 67127 94483 49643
73559 80833 46332 95928 38140 15780 31820 55197 82170
27392 06310 97100 62603 83262 54290 00440 72533 19613
77965 52746 43905 17609 40430 08237 56411 50129 81796
01830 28081 01097 87809 02441 73368 09777 14813 54343
87525 46136 37567 51399 15776.
```

Yet, by a proof as old as Euclid, we know that this number, like the sixteen other perfect numbers, is the sum of all its divisors except itself—just as surely as we know that $6 = 1 + 2 + 3$.

But how many perfect numbers are there? Is the number finite or infinite?

Euclid's question is still unanswered.

OLD FAVORITES

Not quite so old as perfect numbers, but quite old, are the *amicable numbers*. These are pairs of numbers such that each is the sum of the divisors, including one, of the other. Today many pairs of these numbers are known. (Euler published at one time a list of sixty-four pairs, two of which turned out to be false.) But the ancients knew only one, a pair of numbers which they considered the symbol of perfect harmony. One member of the pair is 220, and the reader may like to see if he can determine the other.

ANSWER

The divisors of 220 are 1, 2, 4, 5, 10, 11, 20, 22, 44, 55, and 110, and these add up to 284, the divisors of which add up to 220.

Although additional pairs of amicable numbers are still being announced from time to time, it has been proved that there are only five pairs the smaller number of which is less than 6,233. The smaller number of each of the other four pairs is printed below and the reader may try his skill to determine the other member of the pair.

| 1184 | 2620 | 5020 | 6232 |

ANSWERS

| 1210 | 2924 | 5564 | 6368 |

THE NUMBER SEVEN HAS BEEN HELD IN esteem since antiquity as being the unique number among the first ten. It holds this title because it is the only one which is not produced by any of the others, with the exception of one, of course, and which does not produce any other: six, eight, nine and ten being produced by the primes two, three, five and seven, and all being produced by one, the unit.

"On which account," one ancient philosopher concludes, "other philosophers liken this number to Victory, who had no mother, and to the virgin goddess, whom the fable asserts to have sprung from the head of Jupiter: and the Pythagoreans compare it to the ruler of all things."

If he had been less of a numerologist and more of a mathematician, he might have pointed out more significant ways in which seven is unique among the first ten numbers. Seven, for instance, is the only prime among the digits which is not one more than a power of two: two is $2^0 + 1$; three is $2^1 + 1$; five is $2^2 + 1$; but seven is one less than a power of two, $2^3 - 1$. The regular polygon with *seven* sides is the first which cannot be constructed by the traditional methods of straightedge and compass alone.

One of the most interesting dramas in the theory of numbers is the discovery of a relationship between these two apparently unrelated characteristics of the number seven. It is a story that is studded with some of the greatest names in mathematics.

For the beginning, we must go back as always to the Greeks.

To the Greeks, as we have pointed out, numbers were also shapes. Each individual number was thought of as a polygon "with as many angles as units": three, a triangle; four, a square; five, a pentagon; six, a hexagon; seven, a heptagon; and so on. This interest in the shape of numbers extended even to their construction.

The Greeks were especially fond of limiting their constructions to those which were possible with straightedge and compass alone by proved principles of geometry. Their most famous construction problems were the trisection of an angle, the doubling of a cube, and the squaring of a circle. All of these are now known to be impossible with the limitation of straightedge and compass alone. (Even though they have been *proved* impossible it is a rare student who comes to geometry for the first time without trying to make at least one of the constructions and dreaming of achieving mathematical immortality [which he certainly would!] when he should be trying to prove the theorem of Pythagoras.*)

The problem of constructing regular polygons with straightedge and compass alone is somewhat different from these others in that it is possible to construct some polygons, impossible to construct others. The man who

* Without the limitation of straightedge and compass alone, all the constructions are possible.

determined the criterion for such constructibility was well on his way to mathematical immortality with his discovery before he was nineteen years old! But that was long after the Greeks.

With straightedge (a ruler without any marking upon it) and compass alone, the Greeks made a square within a circle by bisecting the angle of the diameter, and an octagon by bisecting in turn the resulting angles. It was obvious to them that continuing in this way they could construct all polygons the sides of which are powers of two. They constructed the triangle and the pentagon in more roundabout ways, first constructing a hexagon and a decagon and then combining pairs of angles. It was also obvious that they could bisect the angles of hexagon and decagon and construct more polygons in this way.

The problem of the constructible polygons was thus reduced to the construction of regular polygons with a prime number of sides. The Greeks constructed two, three, and five. They stopped, defeated, at seven. Were there more constructible polygons? If there were more, was their number finite or infinite? These questions remained unanswered for two thousand years. In that time no one constructed by straightedge and compass alone a regular prime polygon with sides numbering more than five.

The opening act of the drama of the constructible polygons had taken place in Greece before the birth of Christ. Act II, scene 1, was laid in France; scene 2, in Russia. At the time, no one in the audience suspected that the second act was even connected with the first. The leading role in it was played by Pierre Fermat and the role he played was that of a very great mathematician being dead wrong.

Fermat was concerned, not with the constructible polygons, but with a particular form of number that he believed was always prime. In the theory of numbers the search for such a form which will invariably generate primes has been intensive. The only plausible conjecture anyone has made on the subject was made by Fermat. As it happened, his conjecture was false, and the numbers which still bear his name are a permanent reminder of his mistake.

It was the great mathematician's belief that numbers of the form $2^n + 1$ when n is a power of two were, without exception, prime. The first few numbers of the form $2^{2^t} + 1$ are certainly prime.

$$2^{2^0} + 1 = 3$$
$$2^{2^1} + 1 = 5$$
$$2^{2^2} + 1 = 17$$

Fermat himself tested and found prime the next two of the form, 257 and 65,537. These are usually represented by a capital "F" and a subscript having the value of the respective power of two involved as F_3 and F_4. But testing F_5 was beyond even Fermat. In spite of its tidy representation by a capital letter and a single digit subscript, F_5 is a number that runs into the billions.

$$F_5 \text{ or } 2^{2^5} + 1 = 4,294,967,297$$

Fermat made many attempts to find a factor for F_5 ("... j'ai exclu si grande quantité de diviseurs par démonstrations infaillibles," he wrote in 1640) and came to the conclusion (although being a mathematician he never went beyond "I think") that F_5, like the five numbers which precede it, was prime and that all subsequent numbers of the form $2^{2^t} + 1$ were prime. These are those which are now permanently known as the Fermat numbers.

Some might consider the fact that the first five numbers of a particular form are prime a verification that all numbers of that form are prime, especially when the fifth is in the billions. For a mathematician, however, a sampling of any number of numbers is not enough to make a final statement about *all* numbers.

(In sciences other than mathematics a sampling must often serve as verification of a hypothesis. Mathematicians, who by the nature of their science can prove or disprove a hypothesis with complete finality, have a smug little joke that they call "The Physicist's Proof that All Odd Numbers Are Prime." The physicist, so the story goes, starts out by classifying one as prime because it is divisible only by itself and one. Then three is prime, five is prime, seven is prime, nine—divisible by three? well, that's just an exception—eleven is prime, thirteen is prime. Obviously all odd numbers with the exception of nine are prime!)

A mathematician's statement must be proved. To prove his, Fermat would have had to show that every number of the form $2^{2^t} + 1$ must by its very nature be prime. To disprove it, someone had only to show that one number of the form $2^{2^t} + 1$ is divisible by a number other than itself and one.

This is exactly what someone did, but not until almost exactly a century after Fermat made his statement about *"si grande quantité de diviseurs"* which he himself had tried on F_5. This someone was an equally great mathematician—Leonhard Euler, then mathematician at the court of St. Petersburg. Euler, as we have mentioned before, did not like to see mathematical questions lying around unanswered. Did the form $2^{2^t} + 1$ invariably generate primes, as Fermat had conjectured that it did, or didn't it? Answering the question with finality in the negative could be as simple as finding a divisor of F_5, and this is what Euler set out to do.

He first determined that if any Fermat number has a factor, it must be a prime of the form $2^{t+1}k + 1$. A factor of F_5, if such existed, would have to be a number of the form $2^{5+1}k + 1$, or $64k + 1$. With this discovery he greatly simplified the problem of testing the primality of F_5. Only certain possible primes of the form $64k + 1$ needed to be tried. The first few are 193, 257, 449, 577, and 641. As it happens, it was 641 that neatly divided F_5, or 4,294,967,297, and settled for all time that Fermat had been wrong. The form $2^{2^t} + 1$ does not invariably generate primes.

By all rights this should have been the end of the subject. The curtain should have been rung down on the Fermat numbers. But it wasn't. These numbers, whether or not they are prime, are still very interesting. They are obviously, like the powers of two from which they are formed, infinite. Yet in all the infinitude of natural numbers there is not one number which divides *more than one* of the infinitude of Fermat numbers. This means that every one of the Fermat numbers (there being an infinite number of them) has a prime factor which not one of the others has. This fact has been used for a new and neat proof of Euclid's theorem that the number of primes is infinite.†

Mathematicians went right on looking for primes among the Fermat numbers after Euler had shown that they could

† The proof is by George Polya, whose very useful and untechnical little book, *How to Solve It* (Princeton University Press), is highly recommended to the reader.

not possibly all be prime. There was still a mathematically interesting question to be answered. Were there any other primes of the form beyond F_4, or had it been the great mathematician's ill fortune that the only primes in the infinitude of numbers of the form $2^{2^t} + 1$ are the first five?

The third act of the drama took place in Germany, in 1801, with the publication of a small book. Two thousand years after the Greeks and a century and a half after Fermat, it brought to the stage again the five Fermat numbers, this time, to everyone's surprise, in the company of the ancient constructible polygons.

The name of Carl Friedrich Gauss, the very young author of the book, is the only one in the theory of numbers which outshines the other names in this drama. Gauss was one of the three greatest mathematicians who ever lived (Archimedes and Newton being usually named with him); but in the branch of mathematics that is the theory of numbers no name is ever coupled with his. The small book published in 1801, when Gauss was twenty-four, was titled *Disquisitiones Arithmeticae*. Most of the work in it was done between the time when he was eighteen and twenty-one, the most profitable of his many profitable years. The *Disquisitiones Arithmeticae* is credited with systematizing the then completely unsystematized theory of numbers and marking out a path which other, lesser men were to follow gratefully.

It is appropriate, as we shall see, that Gauss took up the ancient problem of the constructible polygons in the seventh section of the *Disquisitiones*. This problem was not one that anybody expected to find in a book on the theory of numbers, for since the time of the Greeks it had always been considered a problem in geometry. When it was solved, however, it was solved by an arithmetician who attacked it with algebra and found the answer in arithmetic.

A detailed exposition of the method by which Gauss solved the problem does not fall within the scope of this book.‡ Briefly, what he did though was to prove that the only lengths which are constructible are those which can be expressed algebraically, using the four basic operations of arithmetic and square roots. These primes, he proved, must be of the form $2^{2^t} + 1$, and none other—in short, the favored primes of Pierre Fermat!

In general then, a regular polygon of n sides can be constructed with straightedge and compass alone only when n is a power of two, or a Fermat prime, or the product of a power of two and distinct Fermat primes.

With the general solution of the problem Gauss had added to the list of the basic constructible polygons just three:

the regular polygon with 17 sides (F_2),
the regular polygon with 257 sides (F_3),
the regular polygon with 65,537 sides (F_4).

Gauss, whose later mathematical achievements were numerous, was always very proud of this one, made when he was just eighteen years old. Supposedly, it was this discovery which decided him between a career in philology and one in mathematics. There is even a story that he suggested a polygon with seventeen sides should be inscribed on his gravestone, as the sphere and circumscribed cylinder, suggesting the formula for the volume of the sphere, decorated Archimedes's. Whether Gauss ever made such a sug-

‡ There is a very fine exposition in Oystein Ore's *Number Theory and Its History* (McGraw-Hill, New York).

gestion, he did point out, after his solution of the problem:

"There is certainly good reason to be astonished that while the division of the circle in 3 and 5 parts having been known already at the time of Euclid, one had added nothing to these discoveries in a period of two thousand years and that all geometers have considered it certain that, except for these divisions and those that may be derived from them . . . , one could not achieve any others by geometric constructions."

There is no seventeen-sided polygon inscribed on Gauss's gravestone, but one does appear on the monument erected to him in his native town of Brunswick.

But even Carl Friedrich Gauss did not answer the question whether the polygon of 65,537 sides (F_4) is the last which is constructible by Greek requirements of the straightedge and compass alone. This is a question that can be answered only when certain questions concerning the Fermat primes are answered. Is F_4 the last prime of the form $2^{2^t} + 1$? If not, and there are more, is their number finite—or infinite?

Since the time of Fermat a great many mathematical man-hours have been expended on these questions. The publication of the *Disquisitiones Arithmeticae,* giving as it did a new significance to the Fermat primes, made the answers even more important. Since Fermat made his conjecture in 1640, not one additional prime after F_4 has been found among the so-called Fermat numbers. At the present time all that has been learned is that fifteen of these numbers are not prime, and therefore not constructible.

Because the list of the known composite Fermat numbers reveals much about the problem, we are printing it below both in the order of the numbers' discovery and in the order of their size. F_{13} (or $2^{8192} + 1$) is the smallest Fermat number the character of which is still unknown. If the reader will recall that we stated in "Six" that the Mersenne number $2^{8191} - 1$ required approximately one hundred hours for testing on a high-speed computing machine, the equivalent of sixty centuries on a desk calculator, he will find the lists printed below rather interesting.

THE FERMAT NUMBERS

Known to be Composite (1954)

In order of discovery		*In order of size*
F_5	in 1732	F_5
F_{12}	in 1877	F_6
F_{23}	in 1878	F_7
F_6	in 1880	F_8
F_{36}	in 1886	F_9
F_{11}	in 1899	F_{10}
F_9	in 1903	F_{11}
F_{18}	" "	F_{12}
F_{38}	" "	F_{15}
F_7	in 1905	F_{16}
F_{73}	in 1906	F_{18}
F_8	in 1909	F_{23}
F_{15}	in 1925	F_{36}
F_{10}	in 1952	F_{38}
F_{16}	in 1953	F_{73}

The reader will note that the character of comparatively large Fermat numbers, such as F_{36} and F_{38}, and the very large F_{73}, a number of really staggering size, is known; yet it is still unknown whether F_{13} is prime or composite. The explanation is simply that the easiest numbers to prove composite are those which have a prime factor among the first primes we would logically try as factors. For this reason it is much easier to factor, say, 14,997 than a smaller number like 8,633. The first prime that will divide 14,997 is three, the first we would try; but the first prime that will divide 8,633 is eighty-nine, the twenty-fourth prime.

F_{73}, the largest known composite Fermat number, is probably the largest number the character of which has ever been investigated. It is so large that if it were printed in the decimal system in standard type and in standard-size volumes, all the libraries of the world could not hold it.|| Fortunately it does not have to be written out in the decimal system before it can be factored. Like all Fermat numbers greater than F_1, F_{73}, if it can be factored, can be factored only by a number of the form $2^{t+2}k + 1$.§ In the case of F_{73} this means a number of the form $2^{75}k + 1$. There are good mathematical reasons for not trying 1, 2, 3, or 4 for values of k, trying instead as the first possible factor of F_{73} the number $2^{75}\cdot 5 + 1$. This *is*, as it turns out, the smallest prime factor of F_{73}. So with a very small amount of work, the composite character of F_{73} was determined half a century ago. The character of F_{13} is still unknown.

It is not likely that the character of F_{13} will be determined in the near future. Nor does it seem likely that the general question of whether the Fermat primes are finite will be

|| We have taken our estimate of the size of F_{73} from W. W. Rouse Ball's *Mathematical Recreations and Essays* (Macmillan, New York), a classic in its field.

§ This is an improvement over the $2^{t+1}k + 1$ mentioned in connection with the factorization of F_5, since it eliminates even more primes as possible factors.

determined soon. Apparently the mathematical tools that are needed have not yet been forged.

Even Carl Friedrich Gauss, greatest though he was of arithmeticians, did not completely solve the problem of the constructible polygons which the ancient Greeks had proposed. He was able to tell us that the regular polygon with seven sides is the first which cannot be constructed, but he was not able to tell us that the polygon with 65,537 sides is the last.

So the story of the constructible polygons and of the Fermat primes does not end, but merely stops. It is a story studded with some of the greatest names in mathematics, but there is still a place in it for another name.

A CHALLENGE

The Mersenne and the Fermat numbers have much in common besides the fact that they both bear the names of men who guessed wrong. Mersenne numbers are of the general form $2^n - 1$; Fermat numbers, of the form $2^n + 1$. Each form will produce primes only for certain limited values of n, and not always for these. For all other n's it is not even necessary to write a number out to show that it can be factored. The problem stated below will make the challenge of both Mersenne and Fermat numbers more understandable.

Problem: For any positive integer s, $x^s - 1$ is algebraically divisible by $x - 1$. Similarly, if s is odd, $x^s + 1$ is divisible by $x + 1$.

$$x^2 - 1 = (x - 1)(x + 1)$$
$$x^3 - 1 = (x - 1)(x^2 + x + 1)$$
$$x^3 + 1 = (x + 1)(x^2 - x + 1)$$

It follows that $2^{rs} - 1 = (2^r)^s - 1$ is divisible by $2^r - 1$ for all s, and that $2^{rs} + 1 = (2^r)^s + 1$ is divisible by $2^r + 1$ if s is odd. The following are special cases of the identities written above.

$$255 = 2^8 - 1 = (2^4 - 1)(2^4 + 1) = 15 \times 17$$
$$511 = 2^9 - 1 = (2^3 - 1)(2^6 + 2^3 + 1) = 7 \times 73$$
$$513 = 2^9 + 1 = (2^3 + 1)(2^6 - 2^3 + 1) = 9 \times 57$$

Using the facts stated above, the reader might enjoy finding some divisors of $2^{12} - 1 = 4095$ and of $2^{12} + 1 = 4097$. He might also consider in what cases the above rule does not give any factors of $2^n - 1$ or of $2^n + 1$ and try to draw a conclusion about when these numbers may be prime.

ANSWER

The number $2^{12} - 1 = 4095$ is divisible by $2^2 - 1 = 3$, $2^3 - 1 = 7$, $2^4 - 1 = 15 = 3 \times 5$, and $2^6 - 1 = 63 = 3^2 \times 7$. As a matter of fact, $4095 = 3^2 \times 5 \times 7 \times 13$.

The number $2^{12} + 1 = 4097$ is divisible by $2^4 + 1 = 17$. Actually, $4097 = 17 \times 241$.

In general, we can find a divisor, other than itself and one, of $2^n - 1$ unless n is prime, and of $2^n + 1$ unless n is a power of 2. Hence, Mersenne numbers are numbers of the form $2^n - 1$ where n is prime; Fermat numbers are numbers of the form $2^n + 1$ where n is a power of 2. As we have seen in "Six" and "Seven," even with these limitations they are not always prime. That is why they present so great a challenge to the mathematician until he can prove whether they are finite or infinite.

109

THE MOST INTERESTING THING ABOUT THE number eight is that it is a cube ($2 \times 2 \times 2$), and the cubes are interesting and tough numbers. Since the time of the Greeks, who gave them their solid 3-D name, these numbers, which are the products of triple multiplication of the same number, have furnished the higher arithmetic with some of its most difficult problems. None has equalled in difficulty the problem which is today very simply the problem of the cubes. In its history, the number eight, in addition to being itself a cube, has been a very significant number.

There are two questions that are usually asked about any group of numbers, and these have of course been asked about the cubes:

> How can the cubes be generally represented in the terms of the other natural numbers?

> How can the natural numbers be represented in the terms of cubes?

One answer to the first question dates from the early Christian era. It is usually credited to Nicomachus, whose *Introductio Arithmetica* in the first century

A.D. was the first exhaustive work in which arithmetic was treated independently from geometry. Cubical numbers, Nicomachus stated, are always equal to the sum of successive odd numbers and can be represented in this way.

$$1^3 = 1 = 1$$
$$2^3 = 8 = 3 + 5$$
$$3^3 = 27 = 7 + 9 + 11$$
$$4^3 = 64 = 13 + 15 + 17 + 19$$

. . .

This first question about the cubes was easy to find an answer to. (There may of course be other answers.) The second question, of the general representation of the natural numbers in terms of cubes, was very difficult; and the answer when found inconsiderately posed a new, different, and much more difficult question about cubes.

When we speak of "representing" one group of numbers in terms of another, we may mean either by multiplication or by addition. It seems natural to think of the primes in terms of multiplication, and the integers are hence generally represented as the product of primes.* On the other hand, it seems natural to think of the cubes, like the squares, in terms of addition; the integers are then represented as the sums of squares, cubes, and the other higher powers.

* When mathematicians begin to think of numbers as the *sum* of primes they get into fantastic difficulties. In 1742 an obscure Russian mathematician named Christian Goldbach offered what is now known as Goldbach's Postulate: that every even number is the sum of two primes. No one doubted this postulate, but it was not until 1931 that a mathematician was able *to prove* that every even number is the sum of not more than three hundred thousand primes! Since then, it has been proved that every sufficiently large odd number is the sum of not more than three primes; hence every sufficiently large even number, of not more than four.

111

Obviously some integers require fewer cubes than others for representation as the sum of cubes. A number which is itself a cube, like eight, requires only one: 2^3. A number like twenty-three, which can be represented only in terms of the three smallest cubes since $3^3 = 27$, requires nine cubes for representation: $2^3 + 2^3 + 1^3 + 1^3 + 1^3 + 1^3 + 1^3 + 1^3 + 1^3$. Eight, however, like twenty-three, can also be said to be the sum of nine cubes since to 2^3 we can add 0^3 eight times for a total of nine cubes.

It is apparent then that *if there is a number that requires the most cubes for representation,* all numbers can be represented by that many cubes with the addition of the necessary 0^3's. But there is no assurance that there is such a number. The requirements for cubical representation might increase as the numbers themselves increase.

There was no serious attempt to answer our second question about the cubes until 1772. In that year a similar question about the squares, after unbelievable difficulty, had at last been answered *with proof*. There is no better example in number theory of the fact that it is easier to state a truth than to prove it. The Four Square Theorem, as it is known, states that every natural number can be represented as the sum of four squares. A little computation with the smaller numbers suggests that this is quite probably true. It is a theorem that is thought to have been familiar to Diophantus. Certainly it was stated by Bachet, the translator through whom Fermat became familiar with the problems of Diophantus. It was then restated as part of a more general theorem and proved by Fermat. (This was the theorem that we met in "Five" to the effect that every number is either triangular or the sum of two or three triangular numbers; *square or the sum of two, three, or four squares;* pentagonal or the sum of two, three, four, or five pentagonal numbers; and so on). Although Fermat remarked that no proof ever gave him more pleasure, as usual the margins of the Bachet Diophantus were too small and the proof died with Fermat. Euler then tackled the problem of proving the part of the theorem pertaining to the squares, devoted in fact forty years, off and on, of his long life to it, without success. Eventually, though, in 1772, with the help of much of the work Euler had already done, the Four Square Theorem was proved by Joseph Louis Lagrange (1736–1813), the man Napoleon called "the lofty pyramid of the mathematical sciences." A few years later Euler brought forth a more simple and elegant proof than Lagrange's of the theorem which had caused him so much difficulty, and it is now the proof generally followed.

With a history like this behind its "twin" in the squares, it did not seem likely that the question of how many cubes are necessary and sufficient to represent any number as the sum of cubes would be easy to answer.

In addition to being the year of the long-sought proof of the Four Square Theorem, the year 1772 offered another incentive for trying to answer the question of cubical representation of numbers. An English mathematician named Edward Waring (1734–1798) had proposed, without proof, a theorem which went on—and on—from where the Four Square Theorem left off. *Every number,* Waring suggested, *can be expressed as the sum of four squares, nine cubes, nineteen biquadrates, and so on through an infinitude of higher powers.* We met Waring in "Three" as the Cambridge professor who published John Wilson's unproved test for primality. Waring was something of a prodigy, being appointed to the faculty at Cambridge before he had obtained the necessary M.A. degree, so that it had to be given to him by royal mandate. During his lifetime he was described as "one of the strongest compounds of vanity and modesty which the human character exhibits." ("The former, however," the writer added, "is his predominant feature.")

We will not at the moment go into the history of Waring's general theorem. It was Waring's good luck, not his fault, that it turned out to be "one of those problems that has started epochs in mathematics." (The words are E. T. Bell's.) As "Waring's Problem" the theorem paid off with an immortality in mathematics which Waring the mathematician never earned.†

At this point we are concerned less with the general theorem than with Waring's choice of nine as the number of cubes necessary and sufficient to represent every number as the sum of cubes. That nine was quite probably the correct choice could have been suggested to him by a little paper and pencil work. If we start out to represent every number as the sum of cubes, we will find by the time we reach one hundred that apparently none requires more than nine and only twenty-three, as we have already mentioned, requires as many as nine. If we continue past one hundred we will find that there is not another number after twenty-three which requires as many as nine cubes until we get to 239.

It was probably on just such paper and pencil work that Waring based his statement that every number can be represented as the sum of nine cubes. It was a good guess, but nothing more. There is no assurance implicit in such

† Oddly enough, in the summary of Waring's life in the *Dictionary of National Biography*, "Waring's Problem" is not mentioned.

paper work, no matter how far into the numbers it is continued, that there are not numbers which require more than nine cubes for representation even though we never find them. No more is there any assurance that the number of cubes required does not tend toward infinity as the numbers themselves do. (This is what happens when we try to represent all numbers as the sum of powers of two, there being no fixed number of powers of two which will be sufficient for representation of all numbers.)

The first step in proving the portion of Waring's theorem which deals with the cubes was proving that there actually exists a finite number of cubes by which every number can be represented; in short, that the number of cubes required does not tend toward infinity. The mathematical symbol selected for this finite number of cubes, if such there was, was $g(3)$. By implication Waring had stated that such a $g(3)$ existed and was nine; that $g(4)$, the finite number of biquadrates necessary for representation of all numbers, was nineteen; and that $g(2)$ was four. Unless a g existed for each power, Waring's theorem had no meaning.

There was no need to prove the existence of $g(2)$, since Lagrange had already proved (by proving the Four Square Theorem) that $g(2) = 4$. But it was not until 1895, more than a century after the publication of Waring's problem, that even the existence, let alone the value, of $g(3)$ was established. At that time it was proved that every number can be represented as the sum of seventeen cubes. This meant that seventeen is sufficient; that the number of cubes required to represent any number, no matter what its size, can never be more than seventeen. Although it had not been proved that *the smallest possible number* of cubes necessary to represent all numbers was seventeen, it had been proved that seventeen was a bound to the number necessary: in short, an estimated value for a finite $g(3)$.

For the next sixteen years mathematicians whittled away at seventeen, reducing the number of cubes sufficient and necessary for representation of every number as the sum of cubes from seventeen to sixteen to fifteen . . . until, at last, it was proved that $g(3)$ was equal to or less than nine. Since it was already known by the representation of actual numbers that at least two (23 and 239) require nine cubes, $g(3)$ is obviously *equal to* nine. This conclusion was reached exactly one hundred and thirty-nine years after Waring had stated that every number can be represented as the sum of nine cubes.

One not familiar with the problems posed by the numbers might be inclined to think it a testimonial to Waring's brilliance in recognizing intuitively what it took his fellow mathematicians well over a century to come to by investigation and proof. This is not the case. For one of the characteristics of the natural numbers—perhaps their most interesting—is that some of the easily guessed relationships among them are the most difficult to prove.

G. H. Hardy, the English mathematician who devoted a great deal of his time to Waring's problem, commented to this effect once as follows:

"No very laborious computations would be necessary to lead Waring to a highly plausible speculation, which is all I take his contribution to the theory to be; and in the theory of numbers it is singularly easy to speculate, though often terribly difficult to prove; and it is only the proof that counts." ‡

‡ For the history of Waring's Problem I am much indebted to Hardy's short paper *Some Famous Problems of the Theory of Num-*

116

It was in 1909, with the value of $g(3)$ at last definitely established as nine, that the problem of the cubes, which had been difficult enough to take more than a century to solve, became an incomparably more difficult—and more interesting problem. In this year it was proved that the numbers requiring as many as nine cubes for representation are finite. Perhaps, as is generally suspected, 23 and 239 are *the only ones* in all the infinitude of numbers.

What is the significance of the fact that only a finite number of numbers require as many as nine cubes? It is that there is some *last* number which requires nine cubes. From that number on, eight cubes are sufficient to represent *all* numbers.

We quote again from Hardy:

"Let us assume (as is no doubt true) that the only numbers which require 9 cubes for their expression are 23 and 239. This is a very curious fact which should be interesting to any genuine arithmetician; for it ought to be true of an arithmetician that, as has been said of Mr. Ramanujan, and in his case at any rate with absolute truth, that 'every positive integer is one of his personal friends.' § But it would be

bers and in *Particular Waring's Problem* (Clarendon Press, Oxford); and, for more recent developments, to Mrs. D. H. Lehmer. Hardy is one of the most quotable of modern mathematicians, and it is only by a determined effort that we have refrained from quoting him even more often than we have. The reader is recommended to his little book *A Mathematician's Apology* (Cambridge University Press).

§ Srinivasa Ramanujan, the brilliant young Indian mathematician who died in 1920 at the age of thirty-two, has a colorful story which must certainly be included in a book on the interesting numbers. He was virtually self-taught until, as a clerk in the government service, he was discovered and brought to England by Hardy. For a few short years, the two men, Englishman and Indian, collaborated on some brilliant mathematical work. It is in the very fine memoir that introduces Ramanujan's collected papers that Hardy absurd to pretend that it is one of the profounder truths of higher arithmetic; it is nothing more than an entertaining arithmetical fluke. It is . . . 8 and not . . . 9 that is the profoundly interesting number."

With the new concept of the number of cubes which would be sufficient to represent all numbers from some point on (perhaps from 239 on), it was necessary to invent a new mathematical expression. Now $g(3)$, the number of cubes necessary to represent all numbers as the sum of cubes, was joined by $G(3)$, the number of cubes necessary to represent all numbers *with a finite number of exceptions,* perhaps only 23 and 239. It had already been established that $g(3)$ is nine; and since the numbers requiring as many as nine had been proved finite, it followed that $G(3)$ must be equal to or less than eight.

This distinction between "Little Gee" and "Big Gee," as they are sometimes called informally, was discovered through work on that phase of Waring's Problem that dealt with third powers, but it had important implications for all other phases of the problem. The existence of $g(s)$ implies the existence of a $G(s)$, and the existence of $G(s)$ implies the existence of a $g(s)$. As a result, the mathematicians now found themselves with two problems for every one that they had had before: to determine a value for "Little Gee" for every power and another, the same or smaller, for every "Big Gee."

(The problem of "Big Gee" had never come up with

---

tells how, visiting his sick friend one day, he remarked that the number of the cab he had arrived in was 1729, "not a very interesting number." Ramanujan replied promptly that on the contrary it was a very interesting number, being the smallest which can be represented as the sum of two cubes in two different ways $(1729 = 10^3 + 9^3 = 12^3 + 1^3)$.

the squares, for $g(2)$ and $G(2)$ are both four. Although all numbers except those of the form $4^m(8n + 7)$ require only three squares for representation, numbers of this form are obviously infinite. There is, therefore, no number at which we can say, "From this point on all numbers can be represented as the sum of three squares.")

The problem of the cubes, as Waring had proposed it, had been solved in 1909; but as so often happens in the theory of numbers the solution of one problem produced another problem.

As they had whittled away at $g(3) \leqq 17$, mathematicians now began to whittle away at $G(3) \leqq 8$. When tables of the actual cubical representation of the numbers up to 40,000 were examined, a curious fact emerged. There are among these only fifteen numbers which require as many as eight cubes for representation; seven is sufficient for all of the others (except of course 23 and 239, which we have already mentioned as requiring nine). The largest number requiring as many as eight cubes is 454. Between 454 and 40,000 *there are no other numbers requiring eight.*

Such paper and pencil work served to indicate, as it had before in the history of Waring's Problem, a point of attack. Mathematicians set out to prove that the numbers requiring as many as eight cubes, like those requiring as many as nine, are finite. When they proved this—as they eventually did, but only recently—the value of $G(3)$ was established as equal to or less than seven. This is where it stands at the date of writing. Yet there are indications in the same paper and pencil work that seven is quite probably not the final answer to the question.

In the table of numbers up to 40,000, there are only 121 for which as many as seven cubes are needed. The

119

largest of these is 8,042. Between 8,042 and 40,000 *there are no numbers that require more than six cubes*. It is generally thought that from 8,042 on there are no numbers which require more than six cubes: that the value of $G(3)$ is probably equal to or less than six.

This is conjectured, not proved.

Yet when someone does prove, as eventually someone quite probably will, that $G(3) \leqq 6$, there is every indication in the table of numbers that this will not be the end of the matter either. Thousand by thousand, those numbers requiring as many as six cubes for representation become rarer. There are 202 numbers in the first thousand numbers requiring six. In the thousand numbers preceding a million, there is only one!

Someone may eventually prove that the numbers requiring as many as six cubes are also finite, like those requiring as many as nine, eight, and perhaps seven, respectively. Then the value of $G(3)$ will have narrowed to four or five, it already having been proved that there are an infinite number of numbers requiring four cubes for representation.

In the tables which have been made, it has been noted that there is a marked tendency for the numbers requiring five cubes to decrease as those requiring four increase. It is possible that eventually the five-cube numbers too will disappear; but if they do, it will be at a point far beyond the ability of man to make tables. This does not matter at all. The exact value of $G(3)$ can never be established by tables; it will have to be *proved* by man.

The problem of the cubes, as it stands today, is one of the most challenging in the theory of numbers. There is no question but that it will be very, very difficult to establish an exact value for $G(3)$. As we said in the opening of this chapter, eight and the other cubes are interesting and *tough* numbers.

ANOTHER PROBLEM OF CUBES

Here is a problem of the cubes which, unlike the one we have been discussing, can be finally solved with a little paper and pencil work. Among all the numbers, there are just four that can be represented by the cubes of their digits; they are, in other words, equal to the sum of the cubes of their digits. What four numbers are they?

ANSWER

$$407 = 4^3 + 0^3 + 7^3$$
$$371 = 3^3 + 7^3 + 1^3$$
$$370 = 3^3 + 7^3 + 0^3$$
$$153 = 1^3 + 5^3 + 3^3$$

A GREAT MANY THINGS ABOUT THE NUM-
ber nine and its relationships with the
other numbers can be expressed by the
equals sign; but there is one property
of nine, known since antiquity and both
interesting and useful, which cannot be
so expressed. This is the fact that
divided into any power of ten, nine always leaves a remainder of one.

When, at the beginning of the nine-
teenth century, a notation very like the
equals sign was at last invented to express this and other similar relationships, all of the numbers took on what
might well be called a mathematical
"new look." No single invention in the
theory of numbers ever posed so many
new and interesting questions. In the
history of the number nine lies the seed
for this sudden flowering.

In the days when computations were
of necessity performed on counting
boards, nine was commonly used as a
check on the computation. Having completed his work and having before him
on the board the beads of his answer,
the computer was curious to know if
the answer was right. Thanks to nine,
there was a very simple way for him to
find out.

He had multiplied, let us say, 49,476
by 15,833 and had obtained the answer

783,353,508. His work no longer remained on the board, only his answer.

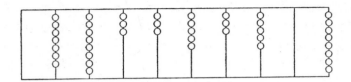

Since he knew that nine leaves a remainder of one when divided into any power of ten and since each bead on the board represented a power of ten, he proceeded to count the beads of the answer as he had previously counted the beads representing the numbers to be multiplied. Today we would add the digits.

$$1 + 5 + 8 + 3 + 3 = 20$$
$$4 + 9 + 4 + 7 + 6 = 30$$
$$7 + 8 + 3 + 3 + 5 + 3 + 5 + 0 + 8 = 42$$

He then divided each of these sums by nine, noting only the remainders.

$$20 \div 9, \text{ a remainder of } 2$$
$$30 \div 9, \text{ a remainder of } 3$$
$$42 \div 9, \text{ a remainder of } 6$$

If the computation is correct, the remainders of the numbers being multiplied will produce the remainder of the product. Since $2 \times 3 = 6$, our computer could go on with a fair amount of confidence to his next problem. There was, however, always the possibility that digits

123

might have been transposed in the answer, a common error which the check will not catch.

This same check can be used for addition and subtraction as well as for multiplication. The sum of the two numbers we multiplied will leave a remainder of five; their difference, a remainder of one. For checking division, we follow the standard rule that the dividend a should equal the divisor b multiplied by the quotient plus the remainder, or $a = b \times q + r$. But instead of using the whole number for this check, we cast out nines and use only the remainders.

$$49476 = 15833 \times 3 + 1977$$

$$\begin{array}{r} 3 \\ 15833 \overline{\smash{)}49476} \\ 47499 \\ \hline 1977 \end{array}$$

or (when the nines are cast out)

$$3 = 2 \times 3 + 6$$

This was the ancient computational check known as "casting out nines." It depends upon the fact we have already noted that nine when divided into one, ten, one hundred, one thousand, or any power of ten leaves a remainder of one. For this reason nine divides a number which is represented in the decimal system only if it also divides the sum of the digits of the number. If it leaves a remainder, this is the same remainder which would be left if the number itself were divided by nine.[*] By this ancient "Rule of Nine" we can say, with only the hesita-

[*] A similar check exists with eleven, which alternately leaves a remainder of $+1$ and -1 when divided into the powers of ten ($+1$ for 1, -1 for 10, $+1$ for 100, -1 for 1000, and so on). To check a computation by elevens, we alternately add and subtract the digits and then divide the total by eleven.

tion of the time that it takes us to add the digits and divide them, that such a number as 9,876,543,210, for example, is divisible by nine. (As it happens, nine divides it exactly 1,097,393,690 times.)

The notation which was at last invented to express a relationship between any two numbers such as that which exists between nine and the powers of ten was a beautifully simple one. It was invented by Carl Friedrich Gauss. The language of his *Disquisitiones Arithmeticae* is Latin, but the real language is that of *congruence*, here used for the first time.

A *congruence* is a relationship, similar enough to the relationship expressed by the equals sign to be very useful, different enough to be very interesting.

$$= \text{ is equal to}$$
$$\equiv \text{ is congruent to}$$

Gauss gave in the *Disquisitiones* the following definition:

Two integers a and b shall be said to be congruent for the modulus m when their difference $a - b$ is divisible by m.

$$a \equiv b \ (\mathrm{mod}\ m) \qquad 5 \equiv 1 \ (\mathrm{mod}\ 2)$$
$$84 \equiv 0 \ (\mathrm{mod}\ 6)$$
$$173 \equiv 8 \ (\mathrm{mod}\ 11)$$

Another way of saying the same thing is this: a leaves a remainder of b when divided by m; 5, a remainder of

1 when divided by 2; 84, a remainder of 0 when divided by 6; 173, a remainder of 8 when divided by 11.

Although the notion of congruence may seem completely unfamiliar when we first meet it, it is not. In fact, it is very familiar. We base every day of our lives on a congruence relationship. When we say that today, for instance, is Tuesday, we are saying that a certain number of days when divided by seven (the week) leaves a remainder of Tuesday.

The day of the week can be very accurately stated as a congruence if we utilize the astronomers' concept of the Julian day. To avoid the confusion which results from months and years of unequal lengths, they number the days consecutively from January 1, 4713 B.C., the beginning of the Julian era. January 1, 1930, which fell on Wednesday, was by this numbering Julian day 2,425,978. With this information and the congruence relationship based on the modulus seven we can compute on which day of the week January 1, 1960, 10,957 days later, will fall.

January 1, 1930 = J.D. 2,425,978 ≡ 2 (mod 7) = Wednesday

January 1, 1960 = J.D. 2,436,935 ≡ 4 (mod 7) = Friday

The general congruence upon which the ancient method of checking a computation by "casting out nines" rests is

$$10^n \equiv 1 \ (\text{mod } 9).$$

This notation tells us at a glance that the difference between one and any power of ten is always divisible by nine. When, instead of looking merely at the powers of ten in relation to the number nine, we look at *all* numbers for the same modulus we find that they fall into nine different groups.

numbers ≡ 0 (mod 9), such as 0, 9, 18, 27, 36, ...
"   ≡ 1 (mod 9), such as 1, 10, 19, 28, 37, ...
"   ≡ 2 (mod 9), such as 2, 11, 20, 29, 38, ...
"   ≡ 3 (mod 9), such as 3, 12, 21, 30, 39, ...
"   ≡ 4 (mod 9), such as 4, 13, 22, 31, 40, ...
"   ≡ 5 (mod 9), such as 5, 14, 23, 32, 41, ...
"   ≡ 6 (mod 9), such as 6, 15, 24, 33, 42, ...
"   ≡ 7 (mod 9), such as 7, 16, 25, 34, 43, ...
"   ≡ 8 (mod 9), such as 8, 17, 26, 35, 44, ...

Every number falls in one of these nine groups, and no number in more than one. With the congruence notation, as in the method of "casting out nines," we are now able to treat all numbers as if they were just nine different numbers. A specially constructed multiplication table gives us all possible products for the modulus nine.

×	0	1	2	3	4	5	6	7	8
0	0	0	0	0	0	0	0	0	0
1	0	1	2	3	4	5	6	7	8
2	0	2	4	6	8	1	3	5	7
3	0	3	6	0	3	6	0	3	6
4	0	4	8	3	7	2	6	1	5
5	0	5	1	6	2	7	3	8	4
6	0	6	3	0	6	3	0	6	3
7	0	7	5	3	1	8	6	4	2
8	0	8	7	6	5	4	3	2	1

Using this table, the reader will find that such apparently dissimilar multiplications as 13×14, 4×32, 22×41, give the same answer (mod 9).

Each pair contains one number $\equiv 4$ (mod 9) and one number $\equiv 5$ (mod 9). The reader will note that 4×5 on the above multiplication table yields 2; and by performing the multiplications indicated, he will find that all three products are $\equiv 2$ (mod 9).

Just as we have looked at all numbers in relation to nine, we can look at them in relation to any n and find that they will accordingly fall into one of n mutually exclusive groups. The best-known way of doing this is according to the number two.

An *even* number is a number $n \equiv 0$ (mod 2).
An *odd* number is a number $n \equiv 1$ (mod 2).

This notation invented by Gauss in the *Disquisitiones* was at once so exact and so easily grasped that many theorems which were already known in other forms were promptly restated as congruences. A case in point is Wilson's Theorem, which we have met earlier in this book. The expression of this theorem as a congruence is today so usual that a mathematician, hearing that the author intended to introduce Wilson's Theorem in "Three" but not mention the congruence notation until "Nine," demanded, honestly disturbed, "But how can you even *state* Wilson's Theorem until you have explained about congruences?" Yet seven years before the inventor of the congruence notation was born, Wilson's Theorem was stated as follows:

128

> If p is a prime, then the quantity
> $$\frac{1 \cdot 2 \cdot 3 \cdot \ldots \cdot (p-1) + 1}{p}$$
> is a whole number.

When young Wilson's teacher, Edward Waring, published this theorem in 1770, he commented: "Theorems of this kind will be very hard to prove, because of the absence of a notation to express prime numbers." It was in connection with this remark that Gauss commented sharply to the effect that mathematical proofs depend on *notions,* not on *notations.* Although today Wilson's Theorem is almost invariably expressed in the notation of congruences as

$$(p-1)! + 1 \equiv 0 \;(\mathrm{mod}\,p),$$

and although the simplest and most direct proof of it (Gauss's own) is based on congruences, the notion remains more important than the notation.

There is, nevertheless, in the history of congruences a strong argument for notations' sharing importance with notions. The type of relationship which is expressed by the three parallel lines of the congruence has been known since the early centuries of the Christian era. There is even an equally brief way of noting it mathematically with the symbol | for *divides.*

To say $m \mid a - b$ is the same as saying $a \equiv b \;(\mathrm{mod}\,m)$.

Yet this long-known type of relationship played no important part in the study of numbers until Gauss found a way of expressing it in a mathematically suggestive form.

129

The three parallel lines of the congruence sign suggest the equals sign and remind us that congruences and equalities, both being equivalence relationships, have certain properties in common. We are familiar with these in equalities:

> If a is any number, then $a = a$.
> If $a = b$, then $b = a$.
> If $a = b$ and $b = c$, then $a = c$.

These properties of the equality relationship are known as reflexivity, symmetry, and transitivity, respectively. All three are also properties of the congruence relationship:

> If a is any number, then $a \equiv a \pmod{m}$.
> If $a \equiv b \pmod{m}$, then $b \equiv a \pmod{m}$.
> If $a \equiv b \pmod{m}$ and $b \equiv c \pmod{m}$,
> then $a \equiv c \pmod{m}$.

The similarities between equalities and congruences which are emphasized by this similar notation suggest that we attempt with congruences certain operations which work with equalities. We have already seen, in the process of "casting out nines," how we can add, subtract, and multiply numerical congruences as we do equations. We can also handle algebraic congruences very much as we handle algebraic equations. The results are usually interesting.

Consider, for instance, a fundamental problem of squares and primes:

> To find a square one less than a multiple of p, where p is a given odd prime.

In the congruence notation this problem is stated more briefly as:

130

Is $x^2 \equiv -1 \pmod{p}$ solvable?

Before we give the general solution of this problem, the reader might like to try to solve it for a few values of p: to find squares which are, respectively, one less than a multiple of the first few odd primes, three, five, seven, eleven, and thirteen. He will find them for only two of these primes, but those he can find very quickly.

Now for the general solution of the problem.

It can be proved, although not here, that the *only* odd primes for which the congruence above is solvable are those which like five and thirteen are of the form $4n + 1$. In the congruence notation

$x^2 \equiv -1 \pmod{p}$ is solvable only when $p \equiv 1 \pmod{4}$.

Closely connected with this problem is a theorem which has the distinction of being the most often proved in the theory of numbers. That it can be approached in so many different ways speaks eloquently for its fundamental importance in number relationships. We mention it here because it is the finest example of that particular type of relationship among the numbers which is brought to the fore by the congruence notation.

The theorem, which is known as the Law of Quadratic Reciprocity, was called by Gauss himself *the gem of arithmetic*. Since at another time Gauss called mathematics *the queen of the sciences* and arithmetic *the queen of mathematics*, this puts the Law of Quadratic Reciprocity at the very pinnacle of science.

The Law of Quadratic Reciprocity was known to mathematicians before Gauss. It was Euler who discovered it, but neither he nor anyone else proved it. Then at the age

131

of eighteen, completely unaware of the work of Euler and others, Gauss rediscovered the law on his own. He found it immediately beautiful, but was not immediately able to prove it. "It tortured me for the whole year and eluded the most strenuous efforts," he wrote. He proved it, at last, in a fittingly beautiful and simple form. At the time he was nineteen years old.

Having proved this gem of arithmetic, Gauss was still so fascinated by it that during the course of his lifetime he composed six more very different proofs. At the time of this writing the total number of proofs of the Law of Quadratic Reciprocity is more than fifty.

The "reciprocity" of the law is that which exists between two different odd primes p and q. The law states that for p and q the two congruences

$$x^2 \equiv q \pmod{p}, \text{ and } x^2 \equiv p \pmod{q}$$

are both solvable or both not solvable unless both p and q are primes of the form $4n - 1$, in which case one of the congruences is solvable and the other is not.

We can see and admire this law in action if we try to determine whether a particular congruence of the type to which the law applies is solvable.

For example:

Is $x^2 \equiv 43 \pmod{97}$ solvable?

This is the same as asking ourselves whether there exists a square which is 43 more than a multiple of 97.

Since not both primes, 43 and 97, are of the form $4n - 1$, we know by the Law of Quadratic Reciprocity that the congruence

132

$$x^2 \equiv 43 \pmod{97}$$

is solvable only if the congruence

$$x^2 \equiv 97 \pmod{43}$$

is also solvable. They stand or fall together: either *both* are solvable, or *both* are not solvable.

To determine the solvability of our second congruence, which will settle the solvability of the first, we proceed, since 97 is greater than 43, to reduce it to

$$x^2 \equiv 11 \pmod{43}.$$

We now have a congruence in which both primes are of the form $4n - 1$. We know by the Law of Quadratic Reciprocity that

$x^2 \equiv 11 \pmod{43}$ is solvable
only if
$x^2 \equiv 43 \pmod{11}$ is *not* solvable.

We now proceed as before, since 43 is greater than 11, to reduce this second congruence. It reduces to a familiar one:

$$x^2 \equiv -1 \pmod{11}.$$

We recognize this as the same problem we met a few pages back: to find a square one less than a multiple of p when p is a given odd prime. We recall from the solution of that problem that the congruence expressed above is solvable only when the prime is of the form $4n + 1$. Since 11 is of the form $4n - 1$, the congruence is not solvable.

We can now work our way back to our original congruence:

Since $x^2 \equiv -1 \pmod{11}$ is not solvable, then by the Law of Quadratic Reciprocity $x^2 \equiv 11 \pmod{43}$ is solvable. Since $x^2 \equiv 11 \pmod{43}$ is solvable, $x^2 \equiv 97 \pmod{43}$ is solvable and therefore by the Law of Quadratic Reciprocity our original congruence $x^2 \equiv 43 \pmod{97}$ is solvable.

We have not actually found *a solution* to the congruence. This is often more difficult than proving that a solution exists, but never so interesting. As it happens, for this particular congruence we can find the numerical solution simply by inspection.

$$x \equiv \pm 25 \pmod{97}$$

This means that any x which differs from a multiple of 97 by 25 will, when squared, be exactly 43 more than a multiple of 97. The lowest positive value for x is 25, and the reader may be interested in testing the original congruence for $x = 25$.

The solution to such a congruence looks quite a bit like the solution to an equation, but there is a significant difference. For an equation, such as $x^2 - 625 = 0$, which also has as its solution ± 25, there are only two values among the infinitude of the integers which when substituted for x will "work." These are $+25$ and -25.

On the other hand, for the congruence which we have just solved, we say also that there are two solutions, but each of these solutions actually stands for an infinitude of numerical values which will satisfy the congruence

$$x^2 \equiv 43 \pmod{97}.$$

In this congruence x can be any number, positive or negative, which differs by 25 from a multiple of 97!

That we are able to speak of this infinitude of values simply as two is indicative of the new perspective on the numbers which we gain from the notion of congruence. Normally when we look at the numbers we try to get as close as possible so that we can see the ways in which they are different from one another. When, however, we look at the numbers in terms of congruences, we move away from them. Suddenly they do not look so much different as they look alike. We are able, as in the solution to our congruence a few pages back, to see an infinitude of numbers as being *the same*. Because they are all congruent to one of a pair of numbers in respect to the same modulus, we can think of them without difficulty, not as an infinitude of different numbers, but as two.

It is a thought-provoking transformation.

For if the numbers, seemingly so regular and predictable as they stretch out by ones to infinity, are capable of such a transformation, what else may they not be capable of?

A LAST PROBLEM

Is the congruence $x^2 \equiv 2 \pmod p$ solvable?

With a knowledge of the solution of this congruence and the solution of the congruence $x^2 \equiv -1 \pmod p$, which we gave in this chapter, combined with the Law of Quadratic Reciprocity, it is possible to determine the solvability of any congruence $x^2 \equiv a \pmod p$.

Although it is not at all easy to prove under what conditions the congruence

$$x^2 \equiv 2 \pmod p$$

135

is solvable, the reader may be able to guess them by actually testing the congruence for the first few squares

$$0, 1, 4, 9, 16, 25, 36, 49, 64, 81$$

and the first few odd primes

$$3, 5, 7, 11, 13, 17, 19, 23, 29, 31.$$

ANSWER

The congruence $x^2 \equiv 2 \pmod{p}$ is solvable when $p \equiv \pm 1 \pmod 8$, so it is solvable for 7, 17, 23, and 31 among the above primes. It is not solvable when $p \equiv \mp 3 \pmod 8$, so it is not solvable for 3, 5, 11, 13, 19, or 29.

IN THIS BOOK WE HAVE OFTEN USED THE symbol of the three dots to indicate that a certain sequence of numbers continues without end. It is appropriate that we stop with the same symbol. For the interesting numbers do not end with nine.

Zero, with which we began our story of the numbers, was the most practical invention in the history of mathematics. The theory of infinite sets, with which we are going to end, may well be the most impractical; yet from the point of view of mathematics, it is incomparably the more important.

Although the modern mathematical theory of the infinite is not properly a part of the theory of numbers, it permeates the modern theory (as it does all of modern mathematics) and develops quite naturally from a consideration of the numbers with which we have been concerned in this book. We have seen in the preceding chapters that the mathematically interesting sequences of numbers are those which continue without end. If the primes were finite, they would be of considerably less interest; and if it is established ultimately that the perfect numbers are finite, their interest will become merely historical. Odd and even numbers, the primes and the composite numbers, the squares, the cubes, the curious pentagonal numbers, all are infinite. These infinite sequences of numbers among the infinite sequence of the natural numbers first suggested the revolutionary idea which is the cornerstone of the modern theory of the infinite.

To understand this idea, we have only to go back to Galileo, who held the cornerstone in his hands but failed to put it into place. In "Four" we told how Galileo pointed out that there are fully "as many" in the infinite set of squares as there are in the infinite set of all numbers. His argument was simplicity itself. Every number, by definition, has a square, which is that number multiplied by itself. We can pair the first square with the first number, the second square with the second number, and so on. We shall never run out of squares until we run out of numbers; and since we shall never run out of numbers, we shall never run out of squares. In a similar way we pair the fingers of the right hand with those of the left, right thumb to left thumb, right forefinger to left forefinger, and so on; when we come out even, we say we have "as many" fingers on one hand as we have on the other. Galileo did no more than to extend this commonly accepted way of determining "as many" to infinite quantities. He pointed out that there is a square for every number throughout the entire sequence of numbers. Squares and numbers can be paired "to infinity." In spite of appearances to the contrary, there are as many squares as there are numbers.

When we say that Galileo did no more than to extend the commonly accepted way of determining "as many" from finite to infinite quantities, we do not intend to minimize his achievement; for in some two thousand years no other mathematician did so much. But Galileo, having come so close to the modern theory of the infinite, did no more. He had showed that logically there are as many squares as there are numbers; then he asked himself the next question. If there are as many squares as there are numbers, can the number of squares be said to be *equal* to the number of numbers?

Well, how was a mathematician going to answer that one?

If there are as many squares as numbers, as he himself had shown, the two sets cannot be said to be unequal. On the other hand, there are obviously many more numbers which are not squares than there are numbers which are squares. At this point Galileo put the cornerstone back on the rock pile, and concluded, as we have seen already in "Four":

"I see no other decision that it may admit, but to say that all Numbers are infinite; Squares are infinite; and that neither is the multitude of Squares less than all Numbers, nor this greater than that; and in conclusion, that the Attributes of Equality, Majority, and Minority have no place in Infinities, but only in terminate quantities."

Three hundred years later, a mathematician named Georg Cantor recognized that inherent in Galileo's definition of "as many" were the concepts of "equality" and of "the same number." To apply to infinities these concepts, usually applied only to finite quantities, he needed a truly precise definition of an infinite set as opposed to a finite set. Such a definition he found in the relationship which Galileo had already perceived between the squares and all the numbers.

An infinite set, defined Cantor, *is one which can be placed in one-to-one correspondence with a proper part of itself.*

This definition obviously does not apply to a finite set. Although we can place *all* the squares in one-to-one correspondence with *all* the numbers, never running out of either squares or numbers, we cannot place the squares less than ten in one-to-one correspondence with the numbers less than ten for the simple reason that we run out of squares before we run out of numbers.

0	0		0	0
1	1		1	1
2	4		2	4
3	9		3	9
4	16		4	
5	25	but	5	
6	36		6	
7	49		7	
8	64		8	
9	81		9	
...	...			

At this point we may well ask ourselves why, if Galileo perceived the essential characteristic of an infinite set as distinguished from a finite set, he did not go on to Cantor's theory of the infinite, three hundred years before Cantor. The answer to this question lies in one of the most ancient axioms of mathematics: the axiom, found in Euclid's *Elements,* to the effect that the whole is greater than the part. Galileo could not bring himself to deny this axiom by saying that the whole (all the numbers) is equal to the part (all the squares). Instead he decided that the Attribute of Equality had no place in infinite quantities. Cantor said in essence: The axiom that the whole is greater than the part *has no place in infinite quantities.*

The mathematical justification for Cantor's revolutionary reversal of Euclid's axiom lies, very simply, in the fact that the reversal *works* with infinite quantities; that is, does not lead us into contradictions, while the axiom that the whole is greater than the part, which works for finite quantities, leads us into contradictions when we apply it to infinite quantities. Before Cantor, mathematicians had struggled in vain to resolve these contradictions. Cantor, defining an infinite set as one which, unlike a finite set, can be placed in one-to-one correspondence with a part of itself, resolved them all by eliminating them.

His theory of the infinite is famous for many *seeming contradictions.* We can prove, for instance, that there are as many points on a line one inch long as there are on a line one mile long; we can prove that there are in all time as many years as there are days.* But we never find

* To prove that there are as many points on the short line as there are on the long, we take the line *AB* and the longer line *CD*, place them parallel to each other, and join the ends *AC* and *BD*. We extend *AC* and *BD* until they intersect at *O*. It is then easy to see that any line drawn from *O* through the two lines *AB* and *CD*

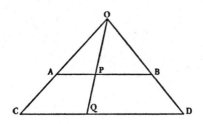

ourselves in the untenable position of having proved in both cases mentioned that the compared sets are equal—and that they are unequal. The justification of consistency—that an axiom does not lead to self-contradictory statements—is all the justification a mathematician needs. By the rules of the game, he is then free to formulate any theory which follows logically from his axiom.

This is exactly what Georg Cantor did. Having defined an infinite set as one which can be placed in one-to-one correspondence with a part of itself, he then defined infinite sets which can be so paired as *equal* and as *having the same number*.

The number, or "power," of an infinite set which can be placed in one-to-one correspondence with the positive integers is the number of *all* the positive integers. It is not the last number, for there is no last number, but the number of *the totality of numbers*. It is a cardinal number because it answers the question *How many?* just as the cardinal numbers we met in "Zero" do, but it is an entirely new kind of cardinal number because it answers the question not about finite, but about infinite quantities. Cantor called it a transfinite cardinal and boldly presented it with a name. As the Greeks had called their numbers by the

---

will intersect them at the points P and Q, respectively. For every point Q on the longer line there will be a point P on the shorter which can be paired in one-to-one correspondence with it.

That there are in all time as many years as there are days is what Bertrand Russell calls the *Tristram Shandy* paradox. Shandy, we recall, spent two years recounting the events of the first two days of his life and bemoaned the fact that at this rate he would fall farther and farther behind in his autobiography. Quite true for a mortal Shandy. But an immortal Shandy, with all eternity at his disposal, would recount the first day's events in the first year, the second day's in the second year, and so on; and eventually he would arrive in his narrative at any given day.

letters in their alphabet, he called his after the first letter of the Hebrew alphabet, *aleph*.

Until this time, infinity, represented by three dots at the end of a sequence of numbers or by the symbol ∞, had been the ultimate in unfinished business: an ever-increasing finite quantity—add one and you always got a larger quantity, a larger number—there was no last number. Of course, Cantor actually changed none of this. The aleph of the positive integers is no last number. It is, for one thing, not a positive integer at all. The relation of the transfinite cardinal of the positive integers to the integers themselves is somewhat similar to the relation of the number one to the fractions. One is not itself a fraction; it is the limit that the fractions approach. No matter how large a fraction we choose (that is, how close to one in value), there is always another fraction which is larger than the first and, therefore, closer to one; yet there is no fraction which exceeds or equals one. In very much the same way this particular aleph, which is not a positive integer itself, is the limit of *all* the positive integers. No matter how large the integer we choose, there is always another which is larger, although it is not any closer to the limit. The essential difference, for this example, between the number one and the number of the positive integers is that while the fractions literally approach the limit one, the positive integers approach their transfinite cardinal only because the larger they get the farther they get from zero. No matter how large the integer we choose, we never get any "closer" to infinity because between us and infinity is always an infinity of numbers equal to the infinity of positive integers.

The idea of infinity, not as something which is in the

process of becoming, but as something which exists—a number which can be handled in many ways just like a finite number: added, multiplied, raised to a power—was as revolutionary as any idea that has ever flowered in the mind of man. Like all revolutionary ideas, it was opposed with emotion, much of it blind and bitter. Even the great Gauss, who thought far ahead of his time and disembarked on many mathematical shores long before their official discoverers, could not accept the idea of a *consummated* infinite.

Probably no great mathematician ever stood more completely alone with his idea than Georg Cantor, but he stood firm:

"I was logically forced, almost against my will, because in opposition to traditions which had become valued by me, in the course of scientific researches, extending over many years, to the thought of considering the infinitely great, not merely in the form of the unlimitedly increasing . . . but also to fix it mathematically by numbers in the definite form of a 'completed infinite.' I do not believe, then, that any reasons can be urged against it which I am unable to combat."

Cantor's confidence lay not only in his mathematics but in mathematics itself. He was always aware of the inherent freedom of mathematical thought, and at another time wrote:

" . . . mathematics is, in its development, quite free, and only subject to the self-evident condition that its conceptions are both free from contradiction in themselves and stand in fixed relations, arranged by definitions, to previously formed and tested conceptions. In particular, in the introduction of new numbers, it is only obligatory to give such definitions of them as will afford them such a definiteness, and, under certain circumstances, such a relation to the older numbers, as permits them to be distinguished from one another in given cases. As soon as a number satisfies all these conditions, it can and must be considered as existent and real in mathematics."

Cantor did not fear such freedom. He recognized that the conditions laid down for it were very strict, arbitrary abuse being kept at a minimum. He recognized also that unless a new mathematical conception was mathematically useful, it was abandoned in short order. Both the mathematical soundness of Cantor's conception of the consummated infinite and the mathematical usefulness of his transfinite cardinals have been borne out by time. Even before his death in 1918, his ideas had been quite generally accepted; and the arithmetic of the transfinite cardinals which we shall detail briefly in the next few pages is now as much a part of mathematics as 2×2.

Having defined cardinality for infinite quantities, Cantor proceeded to establish three important facts: (1) the cardinal number of all the positive integers is the smallest transfinite cardinal; (2) for every transfinite cardinal there exists a larger transfinite cardinal; and (3) for every transfinite cardinal there exists a *next* larger transfinite cardinal. The similarity between the totality of transfinite cardinals and the totality of everyday finite cardinals, or the natural numbers, is apparent. There is a first; there is always a next; and there is no last.

All of these transfinite cardinals Cantor called alephs, but to each one he added a subscript which indicates its place in the sequence. The number of all the positive integers, first and smallest of the transfinite cardinals, has zero as its subscript and is called aleph-null. The next largest transfinite cardinal is aleph-one; the next largest, aleph-two; and so on. These alephs, Cantor showed by mathematical proof, include all possible transfinite cardinals—a very important fact, as we shall soon see.

Is there anywhere an everyday example of an infinity which is larger than that represented by the aleph-null: in other words, an infinity that cannot be placed in one-to-one correspondence with the positive integers?

It was Georg Cantor's achievement, not only to produce such an infinity, but also to produce it by a method so simple that a person with no more knowledge of the theory of the infinite than that which we have been able to expound in the few pages of this chapter has no trouble whatsoever in following his proof. To appreciate his achievement, however, we must realize that just as the positive integers can be placed in one-to-one correspondence with one of their subsets, such as the squares or the primes, infinities which include the positive integers themselves as one of their subsets can also be placed in one-to-one correspondence with them.

An example of such an apparently larger infinity which actually has the same cardinal number as the positive integers is the infinity of all rational numbers. The rational numbers, as we recall from "Two," include all those quantities which can be represented by one whole number over another. Since a number over one represents itself, as $2/1 = 2$, the rationals include the positive integers as well as the fractions. Intuition tells us that there are many more rationals than integers, but intuition tells us that there are fewer squares. And intuition is not mathematical proof. If we can count the rational numbers, we can pair a given fraction a/a_1 with 1, a second b/b_1 with 2 . . . so that in a finite amount of time we shall be able, if we wish, to count to any given rational number.

Well, let us begin. But how?

There is no smallest rational number.

There is no next largest rational number.

Given a rational number a/b as "the smallest," we can always get a smaller by adding one to the denominator, $a/(b + 1)$ being smaller than a/b. Given any two rational numbers, a/b and c/d, no matter how close together, we can always produce another which lies between them by adding the numerators and the denominators of both for a new rational number. Between a/b and c/d lies $(a + b)/(c + d)$.

Have we then found an infinity which is larger than the infinity of positive integers, represented by aleph-null? No, we have not. For it is possible to arrange the rational numbers, although not according to increasing size, in such a way that there is a unique rational number to be paired with each of the positive integers and that, given a sufficient but finite amount of time, we can count to any rational we choose.

We begin by arranging all the rational numbers in subsets according to their numerators, omitting all those with common factors since they will already have been included. We now have an infinite number of rows of rational numbers, and it is obvious that the rows can be placed in one-to-one correspondence with the positive integers:

$$1 \longleftrightarrow \quad \tfrac{1}{1} \quad \tfrac{1}{2} \quad \tfrac{1}{3} \quad \tfrac{1}{4} \quad \tfrac{1}{5} \cdots$$
$$2 \longleftrightarrow \quad \tfrac{2}{1} \quad \tfrac{2}{3} \quad \tfrac{2}{5} \quad \tfrac{2}{7} \quad \tfrac{2}{9} \cdots$$
$$3 \longleftrightarrow \quad \tfrac{3}{1} \quad \tfrac{3}{2} \quad \tfrac{3}{4} \quad \tfrac{3}{5} \quad \tfrac{3}{7} \cdots$$

and so on.

But each column also contains an infinite number of rational numbers which can also be placed in one-to-one correspondence with the positive integers:

1	2	3	4	5	6 ...
↕	↕	↕	↕	↕	↕
$\frac{1}{1}$	$\frac{1}{2}$	$\frac{1}{3}$	$\frac{1}{4}$	$\frac{1}{5}$	$\frac{1}{6}$ ...
$\frac{2}{1}$	$\frac{2}{3}$	$\frac{2}{5}$	$\frac{2}{7}$	$\frac{2}{9}$	$\frac{2}{11}$ ...
$\frac{3}{1}$	$\frac{3}{2}$	$\frac{3}{4}$	$\frac{3}{5}$	$\frac{3}{7}$	$\frac{3}{8}$ ...
...	...	...	...	...	...

We have here infinities upon infinities. If we count by rows, we shall never get to the end of the first row and, therefore, never to the beginning of the second row, or the rational number $\frac{2}{1}$. If we count by columns, we shall never get to the end of the first column and, therefore, never to the beginning of the second column, or $\frac{1}{2}$.

Yet there is a way of counting the rationals so that we can place each of them in one-to-one correspondence with a unique positive integer and so that, in a finite time, we can count to any given rational number. We can do this—Georg Cantor showed—by counting the same arrangement of rows and columns *on the diagonal.*

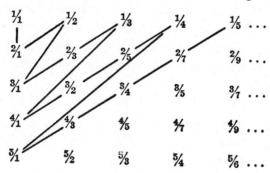

With this method we have a first rational number to be counted ($\frac{1}{1}$) and we always have a next number (in the illustration above, $\frac{2}{1}$). We have no trouble at all in getting to $\frac{2}{1}$ and $\frac{1}{2}$. It is apparent that given sufficient time we can count to any rational number we choose. We shall never run out of numbers to count with. There are, in spite of appearances to the contrary, fully as many positive integers as there are rational numbers.

1	$\frac{1}{1}$
2	$\frac{2}{1}$
3	$\frac{1}{2}$
4	$\frac{3}{1}$
5	$\frac{2}{3}$

and so on.

The transfinite cardinal for both sets is the same, aleph-null.

After a result so contrary to intuition, can we, in accordance with Cantor's proof that for every transfinite cardinal there is a larger transfinite cardinal, place on exhibition such a one greater than aleph-null? Yes, we can. Georg Cantor showed that the infinity of decimal fractions between zero and one cannot be placed in one-to-one correspondence with the positive integers and must, therefore, have a cardinal number greater than aleph-null.

The decimal fractions include both the rational numbers and the irrational numbers, those parts of the unit which, as we saw in "Two," cannot be represented by the ratio of two whole numbers. There are some decimal fractions which at some point, late or soon, terminate in a string of zeros; others which after a certain sequence of numbers begin to repeat that sequence indefinitely; and still others—those which represent the irrational quantities—which by their nature never terminate in zeros and never repeat. All three types can be considered as nonterminating (the string of zeros in the first type continuing to infinity), and all can be represented by the general form

$$0.n_1 n_2 n_3 n_4 n_5 n_6 n_7 n_8 n_9 \ldots$$

where each n represents a given place in the decimal.

Just as it is impossible to write down the first rational number greater than zero, it is also impossible to write down the first decimal fraction; and just as it is impossible to write down the next rational number, it is also impossible to write down the next decimal fraction. Yet it was possible for us to arrange the rational numbers in such a way that there was a first to be counted, and a next, and so on; and we could arrive in a finite length of time at any given rational number. Is there a similar way to arrange the decimal fractions so that they too can be placed in one-to-one correspondence with the integers?

Mathematics allows two ways of answering this question: produce an arrangement, or show that no such arrangement is possible. Cantor did the latter, and did it as simply and smoothly as Euclid two thousand years before him had proved that the number of primes is infinite. We recall (from "Three") that Euclid began by assuming a finite set which included all the primes; he then showed that by multiplying the primes of the set together and adding one, we could always produce a prime not included in the set or a number containing as a factor a prime not

150

in the set. The assumption then that there could be a finite set of all primes was shown false; the primes, infinite.

This is exactly the method Cantor followed. To prove that the set of all decimal fractions between zero and one cannot be placed in one-to-one correspondence with the positive integers, he assumed that by some unspecified arrangement such a correspondence was possible. He assumed a first decimal fraction determined by this arrangement and paired it with the first positive integer. He then assumed a next and paired it with the second positive integer, and so on.

$$1 \longleftrightarrow 0.a_1a_2a_3a_4a_5a_6a_7a_8a_9 \ldots$$
$$2 \longleftrightarrow 0.b_1b_2b_3b_4b_5b_6b_7b_8b_9 \ldots$$
$$3 \longleftrightarrow 0.c_1c_2c_3c_4c_5c_6c_7c_8c_9 \ldots$$
$$\ldots \qquad \ldots$$

He then showed that such an assumption of one-to-one correspondence between the decimals and the integers was false, because he could always produce a decimal fraction which had not been counted. The uncounted decimal fraction he represented as

$$0.m_1m_2m_3m_4m_5m_6m_7m_8m_9 \ldots$$

where m_1 was a digit other than a_1 in the "first" decimal; m_2, a digit other than b_2 in the "second" decimal; m_3, a digit other than c_3 in the "third" decimal; and so on.† A

† Since terminating decimals like 0.25 can be represented as nonterminating decimals in two ways: either as 0.25000 . . . or as 0.24999 . . . , Cantor excluded the digit nine to avoid having the new decimal a different representation of a number which had already in a different form been included in the class of "all" decimals.

decimal fraction formed in this manner could not be included in the assumed arrangement of "all" decimal fractions because it differs from each fraction included in at least one place. The infinity of decimal fractions is thus shown to be greater than the infinity of positive integers because the two sets cannot be placed in one-to-one correspondence; and being greater, its cardinal number must be greater.

This new number Cantor called "the number of the continuum" (the continuum being the totality of numbers such that there is a number for every point on the line). He took as its symbol the German C ($\mathfrak{C}$). This departure from the Hebrew alphabet was extremely significant; for Cantor could not establish where the number of the continuum stood among the alephs. He had shown that aleph-null is the smallest transfinite cardinal and that for every transfinite cardinal there is a next largest (aleph-one, aleph-two, and so on); but, most important, he had shown that all possible transfinite cardinals are included in the sequence of the alephs. Now he had actually produced a transfinite cardinal other than aleph-null, but what aleph was it?

That was the question which Georg Cantor left to the mathematicians who followed him. It is still unanswered. Today the consensus is that the number of the continuum is aleph-one. This is the well-known "continuum hypothesis"; and if it could be established, that is, *proved*, the theory of the infinite would be greatly extended.

Is the number of the continuum aleph-one?

Although it deals with a concept of number and of infinity far beyond the Greeks', the question has all the beguiling simplicity—and all the treachery—of one of their famous questions. Two thousand years ago mathematicians asked, *How many numbers are there which are the sum of all their divisors?* Today they also ask, *Is the number of the continuum aleph-one?* Both questions remain unanswered, but between the asking lie two thousand years under the spell of an apparently simple sequence which begins with zero and continues without end.

Index

INDEX

abacus, *see* counting board
Abel, Niels Henrik, 65
abundant numbers, 85
addition, 7, 124
additive representation of numbers, 22, 111
alephs, 143-153
Alexandria, 44
algebra, 63, 104
amicable numbers, 96
Andrews, F. Emerson, 37
Arabic numerals, 5
Archimedes, 4, 103, 104
Aristotle, 17
arithmetic, 104, 111, 131
Arithmetic of Diophantus, 64, 65, 67

Ball, W. W. Rouse, 107
base-eight system, 38
base-eleven system, 37
base-four system, 38
base-nine system, 37
base-seven system, 37
base-sixteen system, 36-37, 38, 92

base-ten system, 28, 30, 31, 34, 35, 36, 37, 38, 92
base-twelve system, 37
base-two system, 28-40, 53, 92
Bell, E. T., 31, 114
Bhāskara, 11
binary system, *see* base-two system
biquadrates, 113, 115

calendar (*see also* Julian day), 10-11
Cambridge University, 50, 113
Cantor, Georg, 139-153
casting out nines, 122-125, 126, 130
cipher, 5, 6, 43
combinations, 51
composite numbers, *see* prime numbers
computing machines (*see also* National Bureau of Standards' Western Automatic Computer), 28, 34, 37, 53, 85
congruences, 125-136
constructible polygons, 97-109

construction problems (*see also* constructible polygons), 97

continuum hypothesis, 152-153

Coolidge, J. L., 65

counting, 16-17

counting board, 2-3, 4, 5, 122-124

cubes, 29, 110-120, 138

decimal fractions, 149-152

decimal system, *see* base-ten system

deficient numbers, 85

Descartes, René, 87

Dickson, L. E., 20, 84

digits (*see also* zero, one, etc.) 1, 3, 5, 6, 14, 18, 37, 97, 123

Diophantine problems, 64

Diophantus, 63-64, 66, 74, 112

Disquistiones Arithmeticae, 65, 103, 105, 125

divisibility (*see also* Fundamental Theorem of Arithmetic), 20

division, 7-9, 124

doubling of cube, *see* construction problems

duplation, 32

Egyptian mathematics, 1, 2, 62

eight (*see also* base-eight system), 110-120

Elementary Number Theory, 74

Elements of Euclid, 42, 44, 140

eleven, 124

empty set, 13

e pluribus unum, 18

eppur si muove, 69

equalities (*see also* equivalence relationship), 122, 125

equivalence relationship, 130

Eratosthenes (*see also* Sieve of Eratosthenes), 49

Euclid, 4, 24, 42, 44-45, 84, 85, 88, 90, 95, 96, 102, 140, 141, 150

Euler, Leonhard, 75, 77, 88-89, 96, 101-102, 113, 131, 132

even numbers (*see also* even-odd), 18, 43, 111

even-odd (*see also* even numbers, odd numbers), 18, 20, 128

factorial numbers, 51, 52

feminine numbers, 20

Fermat, Pierre, 65-69, 74-75, 87, 112-113

Fermat numbers, 99-109

Fermat's Great Theorem, 66-68

five, 71-82

four, 57-70, 71, 83

Four Square Theorem, 112-113

fractions, 26

France, 99

Fundamental Theorem of Arithmetic, 23-26

Galileo, 59-61, 69, 138-140

Galois, Evariste, 65

Gauss, Carl Friedrich (*see also* congruences, constructible polygons), 65, 67, 103, 125, 131-132, 144

generating function, 77

geometry, 44, 103, 111

Germany, 103

Gesell, Arnold, 15

Goldbach, Christian, 111

Goldbach's Postulate, 111

Great Amateurs of Mathematics, 65

Great Mathematicians, The, 44

Greek mathematics, 1, 2, 4, 17, 20, 25, 57-58, 62, 71, 74, 81, 83, 84, 98-99, 110, 152-153

Hardy, G. H., 4, 43, 116-118

Heaslet, M. A., 74

Hebrews, 83

hexagonal numbers, *see* polygonal numbers

How to Solve It, 102

Illiac, 94

Indian mathematics, 4, 5, 117

infinite descent, 68-69

infinite sets, 43, 61

infinite sets, theory of, 137

infinity, 17, 137-153

Institute for Numerical Analysis, 83, 93

integers, 9, 10, 11, 111

Introductio Arithmetica, 110

Introduction to Number Theory, An, 43

irrational roots, 25, 26, 149-150

Julian day, 126

Lagrange, Joseph Louis, 113, 115

Lehmer, D. H., 92

Lehmer, D. H., Mrs., 117

Leibnitz, Gottfried Wilhelm, 31-32, 34, 35, 51

Lucas, Edouard, 52, 89, 90

Lucas test for primality, 52-53, 91, 92, 93

masculine numbers, 20

Mathematical Discourses and Demonstrations, 60-61

mathematical invective, 8

Mathematical Recreations and Essays, 107

Mathematician's Apology, A, 117

mediation, 32

Men of Mathematics, 31

Mersenne, Marin, 87-88

Mersenne numbers, 87-96, 106, 108-109

multiplication (*see also* peasant multiplication), 7, 30, 123, 127-128

multiplicative representation of numbers, 22, 111

Napoleon, 113

National Bureau of Standards' Western Automatic Computer, 53, 83, 90-96

natural numbers, 11, 13, 14, 17, 19, 26, 43, 58, 59, 71, 75, 77, 116

negative numbers, 9, 10, 11

New Numbers, 37

Newton, Isaac, Sir, 103

Nicomachus, 110-111

nine, 122-135

number, 12, 15, 17, 18

number theory, *see* theory of numbers

Number Theory and Its History, 104

odd numbers (*see also* even-odd), 17, 18, 41, 43, 57-58, 111

one, 12, 15-27, 28, 71-72, 83, 143

one-to-one correspondence, 16, 60-61
onion, 18
Ore, Oystein, 104
Oriental mathematics, 4

pair system, 28-30
Palomar telescope, 90
partitions, 75-81
peasant multiplication, 32-34
pentagonal numbers, 71, 72, 138
perfect numbers, 83-96, 137, 153
permutations, 51
Polya, George, 102
polygonal numbers (*see also* squares, pentagonal numbers), 71-72, 74
polygons, *see* constructible polygons, polygonal numbers
positional arithmetic (*see also* counting board), 2, 3, 6, 37
positive numbers, 9, 10, 11, 117
prime numbers (*see also* Fermat numbers, Mersenne numbers), 20-25, 35, 41-54, 68-69, 84, 86, 97, 111, 129, 130-136, 137, 138, 146, 150, 151
problems, 14, 39-40, 55-56, 70, 82, 96, 108-109
Pythagoras, 4, 62
Pythagorean Theorem, 62, 64, 69, 97
Pythagoreans, 57, 71, 83, 97

Quadratic Reciprocity, Law of, 131-136

Ramanujan, Srinivasa, 117-118
rational numbers, 146-149
reckoning, 4, 85

rectangular numbers, 42
rectilinear numbers, 42
reductio ad absurdam, 24
religion, 18, 32
Robinson, R. M., 90-91
Roman mathematics, 1, 2, 83
Roman numerals, 2, 5
Rule of Nine, 124
Russell, Bertrand, 12, 142
Russia, 99

seven, 97-109
Shipley, Joseph T., 18
Sieve of Eratosthenes, 49-50
sifr, 5, 6, 43
six, 83-96
special cases, 13, 19, 21
square root, 49
squares, 29, 51-70, 119, 130-136, 138-140, 146
squaring of circle, *see* construction problems
subtraction, 7, 124
sunya, 5, 6, 7
SWAC, *see* National Bureau of Standards' Western Automatic Computer

tables of prime factors, 49-50
tests, 6, 26-27
Theorie des Nombres, 89
theory of numbers, 4, 7, 17, 19, 20, 52, 59, 60, 61, 98, 103, 122
thermometer, 9, 10
three, 12, 17, 41-56, 71, 83
triangular numbers, *see* polygonal numbers
trisection of angle, *see* construction problems

Turnbull, H. W., 44
two, 12, 18, 20, 28-40, 41, 83, 128
Two Square Theorem, 68-69

Uhler, Horace S., 95
University of California, 83, 90
University of Illinois, 94
Uspensky, J. V., 74

Virgil, 11

Waring, Edward, 50, 113-114, 129
Waring's Problem, 113-119
Wilson, John, 50, 113
Wilson's Theorem, 50-52, 128-129
Wright, E. M., 43

zero, 1-13, 19, 21, 37, 137
zero, 6
zeroth power, 30-31